Electronic Equi

Electronic Equipment Reliability

J. C. CLULEY

Senior Lecturer in Electronic and Electrical Engineering
University of Birmingham

Second Edition

M

First edition 1974
Second edition 1981
Reprinted 1982, 1983

Published by
THE MACMILLAN PRESS LTD
London and Basingstoke
Companies and representatives
throughout the world

Printed in Hong Kong

ISBN 0 333 32310 6

Contents

3 Reliability Prediction   59

4 Component Failure Data   103

5 Designing for Reliability   135

Bibliography   171

Index   175

Preface

The first detailed studies of electronic equipment reliability were undertaken to improve the performance of communications and navigational systems used by the armed services. The techniques then developed were subsequently refined and applied to equipment used for many other applications where high reliability was of paramount importance, for example in civil airline electronic systems.

Developments in electronic technology during the last two decades have also caused increased interest in reliability. For example the digital computer has evolved from a rare and somewhat unreliable device into an essential management service for most large industrial and commercial concerns, and a central element in many process- and plant-control systems. Furthermore the introduction of solid-state devices and integrated circuits has so reduced the price of electronic equipment that systems of much greater complexity are now available at economic prices. As a result of the growing contributions of these systems to many industrial and commercial activities their reliability and availability have become of great interest to the equipment user, and much more attention is now paid to these qualities in system specifications.

There is thus a need for all engineers concerned with the design, testing or commissioning of electronic systems or the components from which they are assembled to understand the factors that influence reliability, the ways in which it can be measured and specified, and the means whereby it can be improved where necessary.

This book is designed to present such information at a level suitable for students in the final year of degree and diploma courses. This is considered to be the most suitable part of the course in which to introduce the subject of equipment reliability, since any detailed consideration of the topic requires some acquaintance with many aspects of electronic engineering. Among these are analogue and digital signal transmission and processing, circuit and system design, and component characteristics and tolerances. Although the material of the book is not developed to the level generally reached in postgraduate studies, it would be a suitable introduction to the subject, to be followed by a more detailed examination of particular topics.

Many of the calculations in the book involve simple application of the laws of probability and some knowledge of statistics. Since these topics are not included in all engineering mathematics courses a summary of the material is presented in chapter 2. Apart from this the only mathematics required are elementary algebra and simple integration.

Copies of the British Standards mentioned in the text can be obtained from the British Standards Institution, Sales Branch, 101 Pentonville Road, London N1 9ND.

Since the book was first published the development of large scale integrated circuits has significantly changed the way in which much electronic equipment is built. There has also been a marked trend towards the use of digital systems and digital methods of measurement and transmission, and the applications of microprocessors. All of these changes affect the practical aspects of assessing reliability, designing for high reliability and the effective use of redundancy.

In preparing the second edition I have included new sections dealing with these problems and also with the accompanying problem of software reliability and testing.

University of Birmingham   J. C. CLULEY

1 Introduction

1.1 Introduction

The reliability of electronic equipment has received increasing attention during the past twenty years, as apparatus has grown more complicated and has been applied to a great variety of important tasks. Despite its current interest to the practising engineer, the study, design and evaluation of reliable systems as a professional activity has gained only limited recognition. One reason for this is undoubtedly the wide range of topics which concern the reliability engineer. These include mathematical subjects such as probability theory and statistics, practical subjects such as the characteristics of electronic components and devices and the design and construction of electronic equipment, and some physics and chemistry in the study of failure and corrosion. He may also be involved in the environmental testing of components and apparatus under a wide range of conditions. Thus an electronic engineer who is concerned with equipment reliability will encounter a variety of mathematical and scientific problems in addition to those arising from his own particular branch of engineering.

1.2 Historical Survey

Most of the effort devoted to reliability engineering has been in response to particular problems and applications. Consequently, before embarking on a detailed study, we will examine briefly the short history of the subject. Prior to the 1939–45 war relatively little attention was given to the reliability of electronic equipment. Most of the apparatus was simple, and used comparatively few components, so that an acceptable level of reliability could easily be attained. In addition the equipment was used only in a favourable environment, without excessive heat, vibration or humidity.

During the war circumstances changed drastically; equipment was required to stand desert heat, extreme cold, and high humidity, and it became technically much more advanced and more complicated.

As all of these factors increase the chance of a failure, it became essential to study and improve reliability so that the equipment made for the armed services would perform satisfactorily for long periods. Some results of an American survey of wartime equipment revealed the gravity of the problem; they showed that most electronic equipment was operative for only about one-third of the time, and that on average it cost about ten times its original purchase price to repair, and maintain.

A particular weakness was the thermionic valve, so in 1946 the American airlines set up a research group to investigate valve failure and develop a special class of more reliable valves. This was followed by work on other electronic components, and in 1952 an Advisory Group on the Reliability of Electronic Equipment (AGREE) was established by the American Government. Five years later this group published a report on the specification and testing of equipment reliability, which the government adopted. Thereafter most equipment purchased for military use had a reliability clause in the specification and the same procedure was used later by the civil airlines.

The subsequent development of satellites and other space vehicles made increasing demands on electronic equipment reliability and initiated further research and expenditure on this topic.

In Britain a similar programme for the development of special quality valves for the service was initiated. These valves were intended to be more reliable than the normal commercial products and their characteristics were also much more fully specified. Later, transistors were added to the list of special quality devices, and their use was extended to civil applications, such as telephone equipment where high reliability was important.

In the field of consumer electronics a notable development during the last decade has been the large increase in the rental rather than the sale of television receivers. The reliability of the equipment out on hire is of major importance to the rental firm, since it largely determines the firm's profitability. The availability of a large number of receivers, all serviced by the same organisation, enabled full records to be kept of all faults, and revealed a number of shortcomings. The information was passed to the manufacturers for corrective action.

At the present time, the most demanding application is probably the underwater repeaters used with transoceanic submarine cables, which must have a high reliability over a twenty-year period in order to enable the telephone authorities to obtain a reasonable return on their investment.

Under these conditions, the authority must be prepared to pay a relatively high price for components, due to the cost of the required reliability evaluation programme.

1.3 Definition of Reliability

In addition to its qualitative general meaning, the term 'reliability' is now given an exact meaning when applied to engineering devices or systems. It is defined as 'the probability that the system will operate to an agreed level of performance for a specified period, subject to specified environmental conditions'. Thus the reliability of a small computer might be given as 80 per cent over a 200-hour period, with an ambient temperature of 25°C, and no vibration. In order to define the conditions completely, it may also be necessary to specify the maximum variation in the mains voltage and the relative humidity.

It is important that the entire environment should be specified completely, that is the electrical and electromagnetic situation, the temperature and its variations, the climatic conditions such as salt spray, ice formation, dust storm, humidity, and the mechanical conditions such as the frequency and amplitude of vibration. The electrical environment includes the full range of input signals and inter-ference, the variation in supply voltage, and the size of any switching transients, together with the variation in output load if this is relevant. The electromagnetic environment is important if the equipment must operate near other units which generate large electromagnetic fields. For space and nuclear-reactor electronics we may also need to specify the level of radiation, or the total integrated dose to which the equipment may be exposed.

The general definition of reliability given above means that each value quoted for some equipment relates only to the operating period concerned, and the specified level of performance and working con-ditions. In many cases these are not the particular circumstances under which a user may wish to operate the equipment. It is then necessary to estimate the likely performance under these different conditions from the data available. Although certain general rules can be applied, the effect on equipment reliability of a change in the environment is not always predictable, particularly when extrapolating.

Thus, if we are given the reliability of a given type of transistor at an ambient temperature of 70°C and collector dissipations of 150 mW and 300 mW, we could estimate the reliability over the same period at a dissipation of 200 mW fairly confidently. However, an estimate of reliability for a dissipation of 450 mW would be much more liable to error, since this is considerably above the highest dissipation for which information is available.

Although the definition of reliability given above is the usually accepted one, it omits one factor which may be relevant, that is the age of the equipment. In order to operate without failure for a specified period the equipment must be working correctly at the start of the period of observation, but the definition does not distinguish between new equipment which is starting its life, and equipment which has been operated for a considerable time, and repaired when faulty. New equipment generally suffers from an accumulation of small defects which take some time to locate and repair, so that the failure rate of the equipment falls significantly during the initial period of operation, thereafter changing only slowly with time. Most users are concerned with the performance during the main life of the equipment, not the initial few hundred hours of operation. In this situation the reliability may be quoted after a certain initial 'burn-in' period. This indicates the performance which may be expected after the initial settling-down period has elapsed.

1.4   What is Reliability Worth?

Reliability is but one characteristic of an electronic device or system which must be considered when selecting one of a number of alternative designs. From the users' viewpoint, the most rational criterion for deciding which design is best is that of minimum total life cost. This procedure involves estimating the cost of purchasing equipment and initial spare units, the cost of routine maintenance and spare parts for replacement, and in some cases the cost of being unable to operate if the equipment fails without warning and must be withdrawn from service for repair. The total estimated cost for all of these activities during the total life of the equipment is then used as a criterion of value, and the arrangement giving the least total life cost is adopted. Equipment reliability then becomes an important parameter in the design. As it increases, the cost of buying equipment to be held in reserve in case of failure, the cost of spare replacement parts, and the cost of maintenance staff all decrease. However, the cost of design and development, and the initial purchase price all increase very rapidly as more effort is devoted to increasing reliability. There is thus a stage beyond which no economic benefit can be obtained from any increase in reliability.

Figure 1.1 shows the general way in which costs vary with reliability. The initial purchase price reflects the extra design and development costs which are incurred as the reliability is improved and the cost of more reliable (and thus more expensive) components. The slope of the curve will also depend upon the scale of production; as more units are made the design costs can be spread over a greater number, and the total cost per installation decreases.

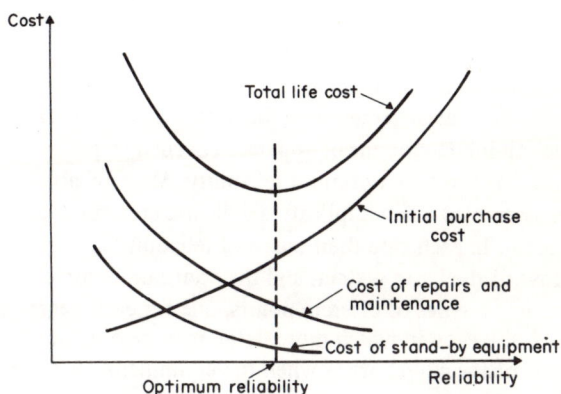

Fig. 1.1 Relation between cost and reliability

In an installation comprising several units, some additional units must be kept in reserve to replace units that are faulty or under repair. As equipment reliability increases, fewer stand-by units are required.

The purchaser of any complex and expensive electronic system should ideally have a large number of different schemes to choose from, with the costs analysed as shown in figure 1.1. He can then estimate the total life cost for each scheme and select the least expensive arrangement. Unfortunately such detailed information is only rarely available and the choice must be based on a few isolated points rather than the complete curves of figure 1.1.

One of the difficulties in collecting the information is that of predicting the precise maintenance costs. Although other users of the equipment, and its manufacturer, should have useful statistics of failure rates and what kinds of faults are likely to develop, the time taken to repair a fault and its cost depend to a great extent upon the kind of repair organisation the user has, and the skill of the staff. Thus the user must consider all of these matters when trying to predict the cost of maintenance, and project this cost many years into the future.

Regardless of the number of items involved, special arrangements may be needed to ensure that spare parts will be available throughout the life of the equipment.

There are, however, some circumstances in which the detailed information required for figure 1.1 can be obtained fairly easily, with sufficient accuracy to choose between alternative proposals. The first of these occurs when most of the designs covered by the curves are redundant systems, in which extra reliability is obtained by using spare or stand-by equipment.

Such an arrangement may be used where the basic system, that is the minimum amount of apparatus needed to perform the allotted task, has less than optimum reliability. More reliable systems can then be assembled using duplicate, triplicate, etc., versions of the basic system. In each case their cost and reliability are simply related to those of the basic system, and the continuous curves of figure 1.1 become a series of discrete points, one for each degree of replication.

A situation in which more detail may be obtained about the initial parts of the curves arises when the equipment has a long development phase. The curves of figure 1.1 may then be plotted, not from a set of different design for the same equipment, but from each successive step in the process of developing the same original design. The development stage would normally end when it became clear that optimum reliability had been achieved, but if time and money are short, it may be necessary to finalise the design at a much earlier stage.

The criterion of minimum total life cost for choosing the optimum level of reliability results in a demand for extremely high reliability for equipment with a long life in which a failure is expensive in repair cost or lost revenue. Examples of such equipment are electronic telephone exchanges, underwater telephone repeaters, and communications satellites.

There are, however, other systems where statutory requirements prescribe a very high level of reliability, since failure may imperil human life. This is the position with the equipment for landing aircraft automatically in bad visibility, or for shutting down nuclear reactors automatically in the event of a failure. In these circumstances the minimum acceptable reliability is fixed by regulation, and the designer may then try to minimise the total life cost subject to this constraint.

As the scope of electronics widens and equipment of greater power and sophistication becomes technically feasible, an increasing number of applications arise in which a major requirement is high reliability. Thus there is continual pressure to develop components which have greater reliability, and system design techniques which can produce more reliable systems from existing components.

1.5 Mean Time between Failures

Although reliability as defined in section 1.3 is generally the most informative index of performance, it suffers from one disadvantage. This arises from the need to specify a particular operating period for the equipment. If the same equipment is operated for a different period, its reliability will be different. Thus one piece of equipment may have a large range of different values of reliability, depending upon the period for which it is used. It is obviously very useful to have some other measure of the performance of the equipment under specified conditions which does not depend upon the operating time. This would facilitate the comparison between different systems with different operating periods. If, for example, we have to choose between two systems with reliabilities of 67 per cent for a 200-hour period and 82 per cent for a 100-hour period the choice requires much more than merely picking the larger of two numbers.

The most useful measure of performance which does not involve the period of observations is the mean time between failure (MTBF).

The MTBF M of a system may be measured by testing it for a total period T, during which N faults occur. Each fault is repaired and the equipment put back on test, the repair time being excluded from the total test time T. The observed MTBF is then given by

$$M = \frac{T}{N}$$

This observed value is not necessarily the true MTBF since the equipment is usually observed for only a sample of its total life. The observed value thus contains a random sampling error, and deductions from the test data must allow for this error.

Another way of expressing equipment reliability is the failure rate; that is, the number of faults per unit time. For many electronic systems the failure rate is approximately constant for much of the working life of the equipment. Where this is the case, the failure rate λ is the reciprocal of the MTBF

$$\lambda = \frac{1}{M}$$

However, where λ changes with time, more than one parameter may be required to express λ as a function of time, and the MTBF may be a more complicated function of λ.

M is usually expressed in hours and the corresponding units of λ are faults per hour. For components, λ is extremely small and the units may be altered to give more convenient numbers. Thus failure rates may be quoted as a percentage per 1 000 hours, failure per 10^6 hours, or failures per 10^9 hours. For example, a system with a MTBF of 2 000 hours has a failure rate of

$$\frac{1}{2\,000} = 0.005 \text{ failures per hour}$$

or

50 per cent per 1 000 hours

or

500 faults per 10^6 hours

The MTBF as defined in the previous section is a concept applicable to any type of equipment which can be repaired by the replacement of a faulty component or unit, and other things being equal, the equipment with the greatest MTBF will be the most reliable, regardless of the period of observation. MTBF thus provides a most convenient index of reliability.

The only qualification necessary concerns the time needed to repair a fault. If this is the same in all cases the equipment with the greatest MTBF will be preferred. However, there may be circumstances in which a short repair time is more important than a long MTBF, and other measures of reliability are needed. These are discussed in section 1.7.

1.6 The Mean Time to Failure

The MTBF is a measure of reliability for repairable equipment; a similar measure is useful for components such as thermionic valves, resistors, capacitors, transistors, etc., which are 'throw-away' items that cannot be repaired. The correct measure for these components is the Mean Time to Failure (MTTF).

This may be calculated from the results of life testing as follows. Let a set of N items be tested until all have failed, the times to failure being $t_1, t_2, t_3, \ldots t_i \ldots t_n$. Then the observed MTTF is given by

$$M = \frac{1}{n} \sum_{i=1}^{n} t_i$$

As with the measurement of MTBF, the life test must be conducted on only a sample of the production batch, so that the observed MTTF will be subject to sampling errors. The failure rate λ will as before be given by

$$\lambda = \frac{1}{M}$$

if λ is independent of time.

For example, if six units were tested until failure, and the times to failure were 320, 250, 380, 290, 310 and 400 hours, the total test time would be 1950 hours, and the MTTF would be

$$M = \frac{1950}{6} = \underline{325 \text{ hours}}$$

1.7 Availability

The manager of an electronic system such as a digital computer, which has a large volume of work to handle each week, is concerned with other factors than the reliability of his equipment. The reliability merely tells him the probability that a job having a certain duration will be processed without a machine failure. Although this is a valuable item of information, the manager needs a further index of performance which takes into account the time lost due to repairing faults. This enables him to decide how much work the computer can handle each week.

If we assume that the computer is switched on for 100 hours per week, the manager needs to know how much of this 100 hours is available for useful work and how much is unavailable as the machine is not working. The relevant index is the 'availability' or the 'up-time ratio' of the system. This is measured from the running log of the machine by dividing all of the 'switched-on' time of the machine in a given period into two categories — 'up-time' U, during which the machine is in

working order, and 'down-time' D, during which the machine is faulty
or being repaired.

The total period of observation is then $U + D$ and we define availability or up-time ratio as

$$A = \frac{U}{U + D}$$

and the unavailability or down-time ratio as

$$B = \frac{D}{U + D}$$

In practice some adjustments to these expressions may be needed.
For example, any time devoted to routine or preventive maintenance is
generally excluded from the calculation, since it is performed outside
the hours of scheduled operation. However, for continuously operating
systems, there is no spare time available for maintenance, and any time
used for this purpose should be counted as down-time.

For an idealised system in which the component failure rate is
constant, the availability has an asymptotic value given by

$$A_\infty = \frac{M}{M + R}$$

where M = MTBF and R = Mean Repair Time.

The initial availability will be greater than A_∞ and will decrease with
time to the steady state value of A_∞.

In practice, however, nearly every complex electronic system exhibits
initial 'teething' or running-in failures which initially increase the component failure rate. Thus the observed availability measured over a
period of some weeks generally rises during the early life of the equipment as the initial faults are corrected.

A further expression for the steady-state availability A may be
obtained in terms of the fault rate $\lambda = 1/M$, and the repair rate $\mu = 1/R$.
Thus

$$A_\infty = \frac{M}{M + R} = \frac{1}{\lambda} \left/ \left(\frac{1}{\lambda} + \frac{1}{\mu} \right) \right.$$

$$= \frac{\mu}{\mu + \lambda}$$

The units in this expression are for μ, repairs per unit time, and for λ faults per unit time.

The availability as defined above can also be interpreted as a probability. If the availability in a given period is A, the probability that at any random instant during the period the system is in working order is also A.

This interpretation is important in the analysis of monitoring or safety systems. These are normally energised continuously, but they are required to function only in the rare event of a plant failure or malfunction. The important measure of performance is then the availability of the safety system, that is, the probability that it still functions correctly at the particular instant at which a plant failure occurs.

1.8   Unavailability or Down-time Ratio

The unavailability of a system may be defined in the same way as the availability; it is the proportion of time that a system is not in working order.

Thus in terms of up-time U and down-time D the unavailability is

$$B = \frac{D}{U + D}$$

where the failure rate is constant, the steady-state unavailability in terms of M and R is

$$B_\infty = \frac{R}{M + R}$$

Finally in terms of the constant rates of repair μ, and failure λ, we have also

$$B_\infty = \frac{\lambda}{\lambda + \mu}$$

Note that by definition $B + A = 1$.

As before, an ideal system with constant failure rate and repair rate will have a transient component of unavailability in addition to the asymptotic value of B_∞. The initial value of B will be below B_∞, the transient component being equal and opposite to the transient component of A.

The value of A is generally the more important of the two, and the complete expression for A shows that the transient term has diminished to about 20 per cent of the asymptotic value A_∞ by a time $t = 4/(\lambda + \mu)$. Since λ is generally much smaller than μ, this time is practically equal to $t = 4/\mu$.

The general variation of A and B with time are shown in figure 1.2. This is, of course, in an ideal situation in which the failure and repair rates are constant.

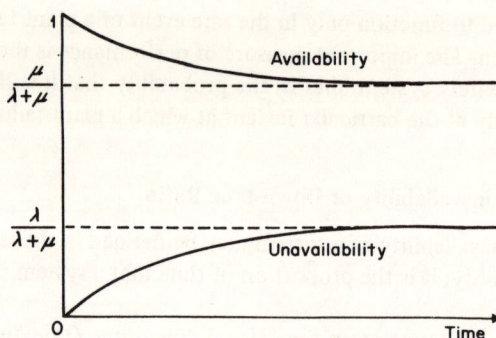

Fig. 1.2 Variation of unavailability and availability with time for system with constant failure rate

In practice the failure rate is generally greater in the initial period, decreasing to an approximately constant level as the initial faults are cleared. Furthermore, the repair rate generally increases during the early stages of operation, as the maintenance staff become familiar with the equipment. For both of these reasons the availability is likely to be less in the initial stages than in the later steady-state phase.

1.9 The Effect of Environment

The definition of reliability requires three categories of information, the test period, the level of performance and the environmental conditions. The last of these includes the complete physical situation in which the equipment is operated, together with its electrical and electromagnetic surroundings, and generally has a major influence on reliability.

The physical factors which must be specified include temperature, acceleration, humidity, atmospheric factors such as rain, snow, duststorms, and ambient pressure, and corrosive conditions such as exposure to sea-water or acid fumes. The most important of these is probably temperature, as all electronic component failure rates increase with temperature. The equipment designer can frequently protect his equipment from part of these stresses by forced cooling, anti-vibration mounts, hermetic sealing, etc., but some of the protective measures may conflict with the electrical requirements of the circuit and some compromise may be needed.

The electrical conditions which must be specified include the variations in power supply voltage (and frequency and waveform for a.c.), the maximum duration and amplitude of any transients which may occur on the power line, and details of any electrical loads. Some equipment may also be susceptible to electromagnetic radiation, and the maximum r.f. field strength at all relevant frequencies must be given. This is particularly important for apparatus which is required to work near high-power radio equipment. A similar specification of the maximum induction field of the environment must be given for equipment which may suffer interference from low-frequency fields. This question arises in the design of low-level input transformers for audio equipment or sensitive a.c. bridges.

Some particular types of equipment may be subjected to unusual conditions which affect reliability, and complete details of these conditions must be included in the reliability specification. For example, some equipment used to monitor nuclear reactor performance is subjected to intense ionising radiation and neutron bombardment. A full description of this facet of the total environment must then be included when specifying reliability.

The various environmental factors to which equipment may be subjected may, of course, occur simultaneously. This complicates the testing of the equipment, since, for example, temperature cycling, humidity, vibration and fluctuation of supply voltage must all be imposed upon the equipment simultaneously. Any attempt to test the equipment subject to only one disturbance at a time will give false and usually very optimistic results for the reliability.

A further complication is that the combination of stresses which produces the greatest failure rate may not apparently be the most severe, and may not be included in a normal test schedule. For example, some types of wire-wound resistor will survive operation at high humidity and high ambient temperature with full electrical loading, but when operated in cool and damp conditions with reduced electrical loading will fail rapidly due to electrolytic corrosion.

This chapter is intended to introduce the main topics concerned in reliability engineering, and to define and explain some of the more important terms used. For completeness, the next section gives a formal definition of the failure distribution function, the relation between reliability and failure rate, and the evaluation of MTBF from the reliability and the failure distribution function. These expressions are used in later chapters which deal with most of the topics introduced above in greater detail.

1.10   Generalised Definitions of Failure Rate and MTBF

The probability of system failure over a given period $F(t)$ is complementary to the system reliability $R(t)$ over the same period, since no other outcome is possible, and these two are mutually exclusive.

Thus

$$F(t) + R(t) = 1$$

and

$$R(t) = 1 - F(t)$$

It is usual to take $t = 0$ as the beginning of the operating period.

An important function derived from $F(t)$ is its derivative with respect to time. Since $F(t)$ is a probability, its derivative is a probability distribution function, defined as

$$f(t) = \frac{\mathrm{d}F(t)}{\mathrm{d}t} = -\frac{\mathrm{d}R(t)}{\mathrm{d}t}$$

whence the probability of a failure during the period from 0 to time t is

$$F(t) = \int_0^t f(t)\,\mathrm{d}t$$

The failure rate can be derived from these expressions by considering a test which starts at time $t = 0$ with n_0 items on test.

After a period, n_s will survive, and $n_f = n_0 - n_s$ will have failed.

The rate at which components fail is thus

$$\frac{\mathrm{d}n_f}{\mathrm{d}t}$$

This may also be interpreted as the number of components which fail in unit time. Since there are n_s items remaining on test, the rate at which each component fails is

$$\lambda = \frac{1}{n_s} \times \frac{dn_f}{dt}$$

But the reliability at time t can be expressed as the probability of no failing in the interval 0 to t. Since n_s or $n_0 - n_f$ items have survived from an initial population of n_0, the reliability may be equated to the proportion of survivors, that is

$$R(t) = \frac{n_s}{n_0} = 1 - \frac{n_f}{n_0}$$

Taking the derivative

$$\frac{dR(t)}{dt} = -\frac{1}{n_0} \times \frac{dn_f}{dt}$$

and

$$\frac{dn_f}{dt} = -n_0 \frac{dR(t)}{dt}$$

Thus we have

$$\lambda = \frac{1}{n_s} \left(-n_0 \times \frac{dR}{dt} \right)$$

Now

$$R(t) = \frac{n_s}{n_0}$$

So we have

$$\lambda = -\frac{1}{R(t)} \times \frac{dR(t)}{dt} \qquad (1.1)$$

This is an entirely general expression, and makes no assumption about the way in which λ varies with time. There is some restriction on $R(t)$, however, since λ must be positive [items having failed cannot be subsequently revived!]. Thus $R(t)$ must be a monotonically decreasing function of t.

The expression for λ may be integrated, giving

$$\int_0^t \lambda \, dt = - \int_1^R \frac{dR(t)}{R} = - \log_e R(t)$$

since $R = 1$ at $t = 0$.

Whence

$$R(t) = \exp\left(- \int_0^t \lambda \, dt\right) \qquad (1.2)$$

We are generally concerned with the simple case in which λ is a constant, so that

$$R(t) = \exp(-\lambda t)$$

In this case

$$f(t) = \lambda \exp(-\lambda t)$$

The method used in statistical theory for finding the mean of a distribution having a density function $f(t)$ is to evaluate the first moment of $f(t)$, which is $t \times f(t)$, and integrate it from $t = 0$ to $t = \infty$. This gives the mean of all the times to failure covered by the distribution, in other words the MTBF, or for non-repairable items the MTTF.

Thus in terms of the distribution function $f(t)$ the general expression for MTBF or MTTF is

$$M = \int_0^\infty t \times f(t) \, dt$$

Now from the definition of $f(t)$

$$f(t) = - \frac{dR(t)}{dt}$$

Thus

$$M = - \int_0^\infty t \times \frac{dR(t)}{dt} \times dt$$

This can be integrated by parts to give

$$M = - [t \times R(t)]_0^\infty + \int\limits_0^\infty R(t)\,dt$$

Now at $t = 0$, $R(t) = 1$, so that $t \times R(t) = 0$ at $t = 0$. As t increases $R(t)$ decreases and a value of k can always be found such that $R(t) < \exp(-kt)$. Thus since

$$\lim_{t \to \infty} t \times \exp(-t) = 0$$

$t \times R(t)$ also tends to zero as t increases.

This gives as a general expression for M, with λ as a function of time

$$M = \int\limits_0^\infty R(t)\,dt \qquad (1.3)$$

For the case of λ constant, this reduces to

$$M = \int\limits_0^\infty \exp(-\lambda t)\,dt$$

$$= -\frac{1}{\lambda} [\exp(-\lambda t)]_0^\infty$$

$$= -\frac{1}{\lambda} [0 - 1]$$

$$= \frac{1}{\lambda}$$

The equations (1.1), (1.2) and (1.3) given in this section are general results which hold for any variation of failure rate λ with time. Although mechanical and electromechanical components may exhibit considerable variations in λ with time, for nearly all calculations relating to electronic systems the failure rate may be taken as approximately constant. This assumption leads to a considerable simplification in the formulae given above, and is supported by arguments presented in chapter 4. In general the simplified expressions for R and M are used in the calculations given in later chapters.

2 The Mathematical Background

2.1 Probability

Since the basic definition of reliability is expressed in terms of probability, the concept of probability, its prediction and measurement are of prime importance. Thus we must first of all define what is meant by probability and give rules for manipulating probabilities, and for calculating probability in various situations.

The definition of probability can be based upon two different principles, the classical approach, or the relative frequency approach. The first of these had its origins in the observations and calculations concerning games of chance which Pascal and Fermat made at the beginning of the seventeenth century. In the following century the ideas were developed and applied to other problems.

2.2 The Classical Definition of Probability

The classical definition of probability is based upon the assumption that the outcome of a random event such as picking a card from a pack can be categorised as one of a number of equally likely results. Then if S is the number of results which can be classed as successful, and there are n possible results in all, the probability of success is defined as

$$P(s) = \frac{S}{n}$$

Thus the probability of drawing a heart in one attempt from a pack of playing cards is found by dividing the number of hearts in the pack

18

by the total number of cards, since picking any one of the thirteen hearts counts as a successful result, and there are as many different possible results as there are cards in the pack.

Thus

$$P(s) = \frac{13}{52} = \frac{1}{4}$$

This calculation assumes that a normal pack of fifty-two cards is used, with a random choice of card, all cards being equally likely to be selected.

The above definition of probability implies limits to the possible values of P, since an event which always occurs is assigned a probability value of one, and an event which never occurs has a probability of zero.

Thus

$$1 \geqslant P \geqslant 0$$

This basis for calculation can be very useful for many circumstances, but it has some disadvantages. First of all it requires that there should be a finite number of possible results from the event. This may cause difficulties where the number of results may not be known in advance, or may come from an infinite set. A more important objection is that the assumption of equal likelihood for all results of the event is one which classical theory has no means of testing.

We may for example examine a pack of cards to ensure that they are all of similar size and height, without distinguishing marks on the back, so that cutting the pack and selecting a card should be a purely random process. However, when we consider the throwing of dice, no visual inspection can detect whether they are loaded. Thus classical theory can assign a probability to the result of throwing ideal, symmetrical dice, but it cannot predict the probability of throwing a particular number with real dice, which may be loaded. The real situation can be handled only by using the 'relative frequency' definitio of probability.

This approach is particularly important in reliability studies, where the reliability of a component cannot be calculated from any 'equally likely outcome' considerations, but must be discovered experimentally.

The classical approach cannot however be dismissed, since many practical situations are in fact very close approximations to the 'equally likely outcome' assumption, and classical theory then gives a simple means of calculating probability.

2.3   The Relative Frequency Definition of Probability

The relative frequency definition of probability is based upon experiment and defines the actual probability as the limiting value of the measured probability, as the number of experiments conducted increases indefinitely. Thus if some event is repeated n times, and m of these are successful, the measured probability of success is $m/n = p_n$. The true probability is then defined as the limiting value of p_n as n increases indefinitely; that is

$$p = \lim_{n \to \infty} p_n$$

This definition has the advantage of depending only on experimental results and requires no assumption about a set of equally likely outcomes. Its disadvantage is that we can never determine any probability exactly, since we cannot perform an infinite number of experiments. We are thus forced to accept some degree of uncertainty in all attempts to determine probability experimentally; we can never say, 'The probability that this transistor will fail during the next year is 0·1 per cent'.

All that tests in a sample of the population can determine is the behaviour of the sample. If we assume that the sample is representative of the complete population, we can say only that there is a certain probability that the parent populations' characteristics lie between certain limits. We could for example deduce from a test that there was a 90 per cent probability that between 0·07 per cent and 0·2 per cent of a batch of devices will fail during the next year. From tests on a given sample we accumulate only a given amount of information, so that the greater the confidence we wish to place in the result, the wider the limits on the performance figures must be.

2.4   Complementary Events

If the outcome of a trial can always be put into one of two possible categories (never in both), the categories are called complementary or mutually exclusive. Success and failure are complementary categories.

Using the classical definition of reliability, if, from $a + b$ equally likely outcomes of a test, a are successful and b are failures, the probability of success

$$P_s = \frac{a}{a + b}$$

and the probability of failure is

$$P_f = \frac{b}{a + b}$$

Thus we see that

$$P_s + P_f = \frac{a + b}{a + b} = 1$$

Given the probability p of some event, the probability of its complement, that is, that the event will not take place is $1 - p$.

Thus if the probability that an equipment will survive for a given period, that is, the reliability is p, the probability that it will fail is

$$p_f = 1 - p$$

This is a useful result as it is often simpler to calculate the probability of failure than the probability of success.

This result can be extended to any number of mutually exclusive categories. If there are n categories, with probabilities $p_1, p_2, p_3, \ldots p_n$, these must satisfy the relation

$$p_1 + p_2 + p_3 + \ldots + p_n = 1 \tag{2.1}$$

2.5 Compound Events

We frequently require to calculate the probability of two or more events occurring, when the probability of each is known. This is easiest to find when the two events are independent; that is, the probability of one having a particular outcome does not depend on the outcome of the other. This is given by the product rule, which is that if p_1 and p_2 are the probabilities of success in the two events, the probability that they both occur is

$$p = p_1 \times p_2 \tag{2.2}$$

This may be extended to more than two events. Thus if p is the probability that a single trial will succeed, the probability that two trials will both succeed (assuming identical circumstances) is p^2.

This product rule is directly applicable to series systems, in which the input of each unit is connected to the output of the previous unit. In order for the complete system to operate correctly, each unit must operate correctly. Thus if the probabilities of success, or in other words the reliabilities of the units are $R_1, R_2, R_3, \ldots R_n$, the probability that they will all operate correctly, that is, that the system will function, is given by the product rule

$$R = R_1 \times R_2 \times R_3 \times \ldots \times R_n \qquad (2.3)$$

for n similar units of reliability R_1, this is $R = (R_1)^n$.

If, for example, a small radio installation consists of three units, (1) receiver, (2) transmitter and aerial, and (3) power unit, with reliabilities for a given operating period of 0·9, 0·85 and 0·8 respectively, the system reliability would be

$$R = 0.9 \times 0.85 \times 0.8$$
$$= \underline{0.612}$$

The probability of a system fault during this period is thus

$$p_f = 1 - R = \underline{0.388}$$

The product rule gives the joint probability that a number of events will all be successful. For some circumstances, however, we require the probability that one or more events will be successful. This is easiest where the events are mutually exclusive. If, for example, we have a box containing eight $0.1\,\mu F$, seven $0.5\,\mu F$ and five $1\mu F$ capacitors, the total number of capacitors is twenty. If we pick a capacitor at random, the probabilities of the three values are

$$p_1 = \frac{8}{20} = 0.4 \qquad (0.1\mu F)$$

$$p_2 = \frac{7}{20} = 0.35 \qquad (0.5\mu F)$$

$$p_3 = \frac{5}{20} = 0.25 \qquad (1.0\mu F)$$

These results are mutually exclusive since a capacitor from the box must have only one of the three values.

If we now calculate the probability of picking either a $0.1\,\mu F$ or a $0.5\,\mu F$ capacitor on the basis of equal likelihood, we have twenty possible capacitors to pick from, and success in $8 + 7 = 15$ cases. Thus the combined probability of picking either value is

$$p_4 = \frac{15}{20} = 0.75$$

Clearly, this is equal to

$$p_1 + p_2 = 0.4 + 0.35$$

In general, if the probability of outcome 1 is p_1, and of the outcome 2 is p_2, the probability that one or other will occur is given by the addition rule

$$p = p_1 + p_2 \qquad (2.4)$$

As before, this is valid only if both events cannot occur together. This result can be extended to any number of mutually exclusive outcomes.

For example, if we have a box containing only resistors of the following values, with the probabilities of picking the value at random given by

Value	Probability
100·0 Ω	0·12
470·0 Ω	0·17
680·0 Ω	0·30
1·0 kΩ	0·11
8·2 kΩ	0·08
15·0 kΩ	0·22

This is a valid set of probabilities since their sum is 1. The probability of selecting at random a resistor of value less than 1 kΩ is the sum of the probabilities of selecting 100 Ω, 470 Ω and 680 Ω, since all results of the selection are mutually exclusive.

Thus

$$p_1 = 0.12 + 0.17 + 0.30 = \underline{0.59}$$

The probability of selecting a resistor of value 1 kΩ or greater is the complement of this; that is

$$p_2 = 1 - 0.59 = \underline{0.41}$$

This is also equal to the sum of the probabilities of selecting 1 kΩ, 8·2 kΩ and 15 kΩ (0·11 + 0·08 + 0·22).

2.6   Joint Probability for Non-exclusive Events

The calculations above are all dependent upon the fact that the result of a selection can be only one value of resistance or capacitance, so that the classes into which the result can be placed are mutually exclusive. We frequently require to analyse a situation in which the results of two events are not mutually exclusive, and both events can have successful outcomes.

A particular case in reliability studies is that of a duplicate system in which two identical channels are provided, but one alone can satisfy the operational requirements. If suitable change-over switching is incorporated, the system will not fail until both units are faulty. Thus to determine the overall reliability we require the probability that either one or both of the channels are working, given the reliabilities R_1 and R_2 of the two channels individually for a prescribed operating period.

Here the two channels are considered independent, in that a fault in one channel makes no difference to the probability of a fault in the other channel. If a channel without a fault is classed as successful, it is possible for both channels, that is, both events, to be successful simultaneously. We can, however, divide the situation into three different successful outcomes and one unsuccessful outcome, all of which are mutually exclusive. The simple addition rule is then applicable.

The unsuccessful outcome is that in which both channels are faulty and the system is inoperative. The three successful situations, which are mutually exclusive, are

(a) both channels working
(b) channel 1 working and channel 2 faulty
(c) channel 2 working and channel 1 faulty

The probabilities of the channels being faulty are

$$U_1 = 1 - R_1 \text{ for channel 1}$$

$$U_2 = 1 - R_2 \text{ for channel 2}$$

The probabilities of the three states can be found by the product rule for joint probability as

$$P_a = R_1 \times R_2$$

$$P_b = R_1 \times U_2 = R_1(1 - R_2)$$

$$P_c = R_2 \times U_1 = R_2(1 - R_1)$$

Thus, the overall probability of one or two channels working is

$$R_s = P_a + P_b + P_c$$
$$= R_1 R_2 + R_1 - R_1 R_2 + R_2 - R_1 R_2$$
$$= R_1 + R_2 - R_1 R_2 \tag{2.5}$$

A simpler derivation of this result comes from considering the complementary situation. The duplicate system fails if both channels fail. This is a combined probability calculated by using the product rule. Thus

$$U_s = U_1 \times U_2$$

whence

$$1 - R_s = (1 - R_1)(1 - R_2)$$
$$= 1 - R_1 - R_2 + R_1 R_2$$

Thus, as before

$$R_s = R_1 + R_2 - R_1 R_2$$

Here U_s is the probability that the duplicate system is faulty, and is the complement of the reliability, that is, $U_s = 1 - R_s$.

If the two channels have identical reliabilities R_1, the combined reliability reduces to

$$R_s = 2R - R^2 \tag{2.6}$$

The reliability of any other parallel arrangement can be found in a similar manner. If, for example, we have a triplicate system with majority voting, the system will operate correctly if any two of the three channels are working. There are then four possible successful states

(a) channels 1, 2, and 3 working
(b) channels 1 and 2 working, channel 3 faulty
(c) channels 2 and 3 working, channel 1 faulty
(d) channels 3 and 1 working, channel 2 faulty

If the channels are identical, with a reliability for a given period of R, the total probability of any one of these four situations occurring is given by the simple addition rule, since they are mutually exclusive. For (a), the product rule gives a probability of R^3, since three channels, each with a probability of correct operation of R, must simultaneously be working. For (b), (c) and (d) two channels must be working, each with a probability of R, and the third channel faulty, with a probability of $(1 - R)$. Thus the joint probability for each state is $R^2(1 - R)$.

The total probability of the system working is then the sum of the four probabilities, that is

$$R_s = R^3 + 3 \times R^2(1 - R)$$
$$= 3R^2 - 2R^3 \tag{2.7}$$

2.7   General Addition Rule

In the general case, if the probabilities of two events are p_1 and p_2, the probability that one or both will occur is

$$p = p_1 + p_2 - p_{1,2} \tag{2.8}$$

where $p_{1,2}$ is the probability of both events occurring simultaneously. If the two events are mutually exclusive, they can never occur simultaneously, so that $p_{1,2} = 0$.

The addition rule then reduces to its simpler form

$$p = p_1 + p_2$$

In addition to its use in evaluating joint probabilities, this approach can be used to calculate numbers of joint events in a given set of trials. The calculation can be assisted by using a diagram similar to the Karnaugh map used to simplify logical expressions, as shown in the following example.

We assume that a batch of experimental electrolytic capacitors has been tested. These can have two faults, high leakage, or low capacitance. Using an automatic component tester, 2·5 per cent of the batch are rejected for high leakage. The entire batch is then tested for low capacitance, and 1·3 per cent are rejected. These rejects are then tested for leakage and 0·5 per cent of the batch are found to have both faults.

Let A denote the characteristic of high leakage, the complement $\bar{A}$ denoting normal leakage, and B denote low capacitance with $\bar{B}$ denoting normal capacitance (a capacitance above the nominal value is not classed here as a fault). Then $A\bar{B}$ denotes a capacitor with high leakage and normal capacitance. The four possible combinations can then be shown diagrammatically as in figure 2.1. Here the classes AB and $A\bar{B}$ together constitute class A. The proportion in class AB is 0·5 per cent, so the proportion in $A\bar{B}$ is 2·0 per cent, since the sum of AB and $A\bar{B}$ is 2·5 per cent. Similarly the proportion in $\bar{A}B$ is 0·8 per cent since the sum of classes $\bar{A}B$ and AB must be 1·3 per cent. The percentage figures shown can then be entered in figure 2.1, and the sum of the three classes AB, $A\bar{B}$ and $\bar{A}B$, all of which denote defects, is $0·5 + 0·8 + 2·0 = 3·3\%$. Thus the fraction of good capacitors is $100 - 3·3 = \underline{96·7\%}$.

The above calculation deals with actual numbers of defects, but the method can be applied in the same way to calculating probabilities. Thus, using the same figures to denote the probability of selecting a capacitor from a given batch with the characteristics given, the joint probability of selecting a capacitor with one fault or the other, or both,

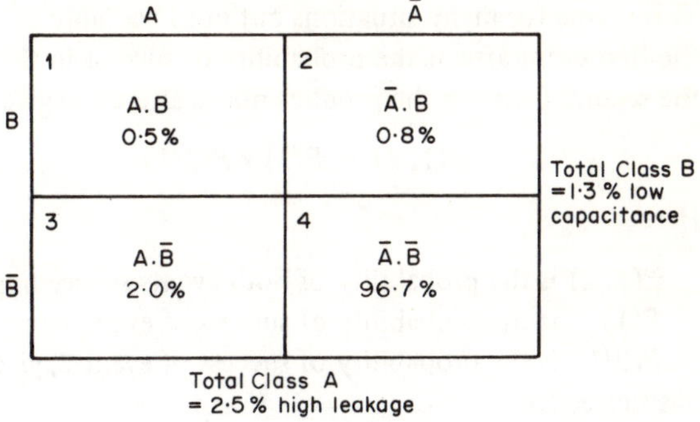

Fig. 2.1 Diagram of capacitor faults

is 3·3 per cent. This also follows from the general addition rule, since

$$P(\text{A or B}) = P_A + P_B - P(\text{A, B})$$
$$= 2\cdot5\% + 1\cdot3\% - 0\cdot5\%$$
$$= \underline{3\cdot3\%}$$

The same argument and diagrammatic method can be applied to tests on components with three or more independent modes of failure.

The method is most useful where the modes of failure are independent, so that they may occur simultaneously in the same component. This is the situation in the example above, since the capacitance and leakage are not dependent upon one another. If the failure modes are mutually exclusive, the calculation is much simpler since only one failure mode can occur in any one component. The simple addition rule

$$P(\text{A or B}) = P(\text{A}) + P(\text{B})$$

can then be applied.

An example of mutually exclusive failure modes for a capacitor would be

A = low capacitance, B = high capacitance

or

A = open circuit, B = short circuit

In neither case can A and B occur together in the same component.

2.8 Conditional Probability

The simple product rule for joint probability is valid only when the success of one event has no effect upon the success of the second event.

This is true for many situations but not invariably so. When success in the first event affects the probability of success in the second event, the second factor in the product rule is altered as given below

$$P(1, 2) = P(1) \times P(2/1) \qquad (2.9)$$

Here

$P(1, 2)$ is the probability of both events succeeding

$P(1)$ is the probability of success of event 1

$P(2/1)$ is the probability of success of event 2, given that event 1 has succeeded.

$P(2/1)$ is called a 'conditional probability' and it will differ from $P(2)$, the probability of success of event 2 considered in isolation, only if event 2 depends upon the result of event 1.

Clearly, we could consider event 2 as the initial event, and obtain the expression

$$P(1, 2) = P(2) \times P(1/2)$$

The way in which this expression can be used is shown in the following example.

Suppose we have a box containing sixty capacitors, each having one of three voltage ratings and one of three capacitance values, the numbers of each type being as shown in table 2.1.

Table 2.1

		Value			
	Voltage rating	A $0.1\,\mu F$	B $0.22\,\mu F$	C $0.47\,\mu F$	Totals
D	63	5	2	3	10
E	125	7	6	8	21
F	250	8	11	10	29
	Totals	20	19	21	60

If we consider a capacitor selected at random, the following probabilities may be assigned to the various events, considering each of the sixty capacitors equally likely to be picked

$P(A)$ = Probability of selecting a $0.1\,\mu F$ capacitor

= 20/60 = 1/3

$P(B)$ = Probability of selecting a $0.22\mu F$ capacitor

= 19/60

$P(D)$ = Probability of selecting a 63 V capacitor

= 10/60 = 1/6

$P(E)$ = Probability of selecting a 125 V capacitor

= 21/60 = 7/20

$P(D/A)$ = Probability of selecting a 63 V capacitor from the $0.1\mu F$ capacitors

= 5/20 = 1/4

$P(E/A)$ = Probability of selecting a 125 V capacitor from the $0.1\mu F$ capacitors

= 7/20

Then from the expressions for conditional probability, the probability of selecting a capacitor of rating 125 V and value $0.1\mu F$ is

$$P(A, E) = P(A) \times P(E/A)$$
$$= 1/3 \times 7/20 = 7/60$$

Similarly we could use the alternative expression

$$P(A, E) = P(E) \times P(A/E)$$
$$= 7/20 \times 1/3 = 7/60$$

In this instance we could divide the possible outcome into the selection of one of nine categories of capacitor, since each of the three voltage ratings can be combined with any one of the three capacitance values. The particular combination we require is 125 V rating and $0.1\mu F$, and since seven of the sixty capacitors belong to this category, the probability of selecting one of them is

$$P(A, E) = 7/60$$

2.9 Bayes' Theorem

In the previous example the joint probability $P(A, E) = P(A) \times P(E/A) = P(E) \times P(A/E)$.

Thus, we may write

$$P(A/E) = \frac{P(A, E)}{P(E)}$$

$$= \frac{P(A) \times P(E/A)}{P(E)} \qquad (2.10)$$

This expression may be written in an alternative form by noting that $P(E)$ can be expressed as the sum of three products

$$P(E) = P(A, E) + P(B, E) + P(C, E)$$

$$= P(A) \times P(E/A) + P(B) \times P(E/B)$$
$$+ P(C) \times P(E/C)$$

$$= \sum_A P(A) \times P(E/A)$$

in the general case.

Thus, we may write as an alternative form

$$P(A/E) = \frac{P(A) \times P(E/A)}{\sum_A P(A) \times P(E/A)} \qquad (2.11)$$

This enables us to use the result of an experiment to modify our estimate of probability. For example, the chance of picking at random a $0.1 \mu F$ capacitor from the group of sixty is $20/60 = 1/3$. However, if we select a capacitor and examine the voltage, our estimate of probability changes. If the voltage rating is found to be 63 V we require $P(A/D)$, the probability that a capacitor has a value of $0.1 \mu F$, given that its voltage rating is 63 V.

From the above expressions, we have

$$P(A/D) = \frac{P(A) \times P(D/A)}{P(A) \times P(D/A) + P(B) \times P(D/B) + P(C) \times P(D/C)}$$

$$= \frac{\dfrac{1}{3} \times \dfrac{5}{20}}{\dfrac{1}{3} \times \dfrac{5}{20} + \dfrac{19}{60} \times \dfrac{2}{19} + \dfrac{21}{60} \times \dfrac{3}{21}}$$

$$= \frac{\dfrac{5}{60}}{\dfrac{5}{60} + \dfrac{2}{60} + \dfrac{3}{60}} = \frac{5}{10} = \frac{1}{2}$$

Thus, our estimate of the probability that the capacitance is $0.1\mu\text{F}$ changes from 1/3 to 1/2 when we are informed that the voltage rating is 63 V.

An important application of Bayes' theorem is in making the best use of data obtained from life tests on equipment, by combining them with prior information obtained from estimates at the design stage and experience with the operation of similar apparatus.

The methods of classical statistics use only data obtained from life tests, usually under simulated working conditions, to estimate, say, the MTBF of some equipment. With the increasingly complex nature of electronic equipment, and its high reliability, life test data is becoming much more expensive to obtain. Furthermore, many customers are demanding higher reliability predicted with greater accuracy.

In a practical situation any particular equipment will have its reliability under continuous assessment as the project proceeds, and as more of the equipment details are decided, more use can be made of operating experience with similar equipment already in service. In consequence, the designer of the equipment will have a fair estimate of its likely reliability, based upon a variety of information, before it goes into production and then on to life test.

It seems unreasonable that all of this information should be completely neglected, as it is in classical statistics, when attempting to evaluate life test data. For this reason Bayesian statistics have been proposed as a method of combining all known information about an equipment in order to get the best possible estimate of its reliability. The method uses all relevant information available about the equipment to form a 'prior' estimate of reliability before testing starts. The estimate is in effect a probability distribution in which the mean is the prior estimate itself, and the variance expresses the degree of confidence in the estimate. The greater the amount of information, the smaller the variance would be expected to be.

The prior estimate is then combined with the test data from life tests to obtain a final or 'posterior' estimate, again in the form of a distribution.

Certain types of distribution lend themselves to this process by making the combination procedure arithmetically simple. A particularly useful distribution is the gamma distribution, whose shape is determined by two parameters α and β.

If the failure rate is λ, this gives for the probability distribution

$$P(\lambda) = \frac{\alpha}{\Gamma(\beta)} (\alpha\lambda)^{\beta-1} \exp(-\alpha\lambda)$$

The gamma function $\Gamma(\beta)$ is equal to the factorial expression $(\beta - 1)!$ when β is a positive integer. For positive values of β, it is defined as

$$\Gamma(\beta) = \int_0^\infty x^{\beta-1} \exp(-x)\,dx$$

The gamma distribution reduces to the exponential distribution when $\beta = 1$.

The mean value of λ is β/α, the prior estimate of failure rate, and its variance is β/α^2.

Thus, the estimate of MTBF is

$$\frac{1}{\lambda} = \frac{\alpha}{\beta}$$

If we now undertake life tests for a total of T hours, and have x failures, the best estimate of MTBF obtained by combining prior knowledge and test results is

$$M = \frac{\alpha + T}{\beta + x}$$

The posterior distribution of λ will also be a gamma distribution, but with parameters α' and β' where

$$\alpha' = \alpha + T$$
$$\beta' = \beta + x$$

The major difficulty with this procedure is in quantifying the prior knowledge in a form which will yield α and β. One method is to ask the question 'What is the probability that the actual value of λ lies between twice and half of its estimated value?' If this probability is known, the variance β/α^2 can be found from tables, and since the estimated value of λ is equal to β/α, we have enough information to calculate α and β and so determine the prior distribution.

2.10   Probability Diagrams

The arrangement of data in table 2.1 illustrates the use of a diagram
to enumerate the total number of possible outcomes of an event, and
class those which are successful. The simplest version is one in which the
probability attached to each square is the same, so that probability can
be estimated simply by counting squares. This would be the case in
table 2.1 if there were only nine capacitors, each having a different
combination of value and voltage rating.

The result of drawing a card at random from a pack of playing cards
can be illustrated diagrammatically as shown in table 2.2.

Table 2.2
Result of drawing a card

Suit \ Value	1	2	3	4	5	6	7	8	9	10	J	Q	K
Clubs						▨							
Diamonds	▨	▨	▨	▨	▨	▨	▨	▨	▨	▨	▨	▨	▨
Hearts						▨							
Spades						▨							

Squares corresponding to $P(6 + D)$ are shaded.

The probability of drawing an ace is found by counting the squares
corresponding to success, and dividing by the total number of squares,
and is thus

$$P_{(A)} = 4/52 = 1/13$$

The probability of drawing a diamond is similarly found as $13/52 = 1/4$.

The probability of drawing a card of value less than 4 is $12/52 = \underline{3/13}$,
since any of the twelve squares in the first three columns is counted as
successful. We may also evaluate joint probability from the diagram.
If we consider the probability of drawing a red queen, only two squares
are classed as successful, the queen of hearts and the queen of diamonds.
Thus the probability of success is $2/52 = \underline{1/26}$.

If we require the probability of picking, say, either a six or a diamond, we have sixteen squares which are classed as successful, as shaded on table 2.2. Thus the probability of success is

$$P(6 + D) = \frac{16}{52} = \frac{4}{13}$$

Using the addition rule given in section 2.7, this can also be considered as the sum of the probabilities of picking a six, and of picking a diamond, less the probability of picking both together, that is

$$P(6 + D) = P(6) + P(D) - P(6, D)$$

The term $P(6, D)$ is simply the probability of selecting the six of diamonds. This event is included twice in the sum $P(6) + P(D)$.
Thus

$$P(6 + D) = \frac{1}{13} + \frac{1}{4} = \frac{1}{52} = \frac{4}{13}$$

as before.

The diagram method of analysis can also be used, as in the previous capacitor example, where the probability of each outcome may differ. Then, instead of counting squares, the sum of the probabilities assigned to each square must be computed, in order to obtain the overall probability of success.

2.11   Binomial Distribution

We have previously considered the outcome of a single event, and the joint outcome of two events. In each case the number of trials is limited to one or two, and there are only a few different possible results. We frequently require to consider a more complicated situation in which many trials are conducted on one system, or a single trial is conducted on each one of many systems. There are many possible outcomes from such a set of trials, and we require to assign a probability to each of them. This is the same question as asking how the total probability of 1·0 should be distributed amongst the possible outcomes.

A simple example is that of tossing a coin. We assume that the coin is tossed n times, the possible results being 0, 1, 2, ... n heads. We require to assign a probability to each of these results. The basis for the calculation is that of equally likely outcomes. If we consider only two tosses, the possible results are

$$
\begin{array}{cc}
\text{H} & \text{H} \\
\text{H} & \text{T} \\
\text{T} & \text{H} \\
\text{T} & \text{T}
\end{array}
$$

Thus the probabilities are

$$0 \text{ Head } \frac{1}{4}$$

$$1 \text{ Head } \frac{2}{4} = \frac{1}{2}$$

$$2 \text{ Heads } \frac{1}{4}$$

It is essential in this kind of analysis to list all cases which are equally likely, otherwise we may assign a wrong probability.

Thus an incorrect calculation would be obtained by considering that there are only three possible results: 0 heads, 1 head, 2 heads, giving the probability of any of these three results as 1/3.

The error here is that the result '1 head' can occur in two ways — head, tail or tail, head. It is thus twice as likely as the other two results which can occur in only one way.

The simplest check against this error is that in tossing a coin n times there are 2^n different possible results, all equally likely. Unless our analysis includes all 2^n cases we have made an error.

The above argument can be applied to more events, for example, in four tosses the possible results are as described in table 2.3.

Table 2.3
Result of tossing a coin four times

No. of heads	0	1	2	3	4
No. of ways result can occur	1	4	6	4	1
Probability of result	$\frac{1}{16}$	$\frac{1}{4}$	$\frac{3}{8}$	$\frac{1}{4}$	$\frac{1}{16}$

The total number of possible results is $2^4 = 16$. For the case of two heads, for example, we require the number of ways in which we can select two heads from four throws. The first head can be any one of the four and the next any one of the remaining three. However, this process will pick each combination twice, so the number of different combinations is

$$^4C_2 = \frac{4 \times 3}{2 \times 1} = 6$$

The binominal probability distribution listed in table 2.3 is shown in a more informative way in figure 2.2 as a bar chart. Other examples of the binominal distribution are shown in figures 2.3 to 2.5. All of these except figure 2.5 are symmetrical since the probability of success

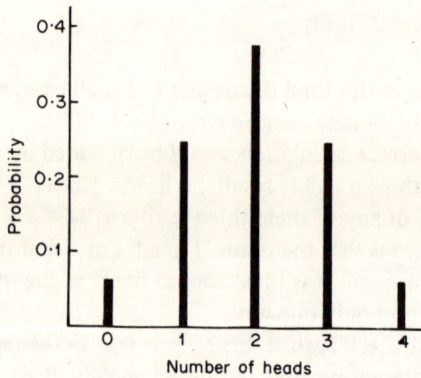

Fig. 2.2 Graphical representation of table 2.3.
Binomial distribution for $n = 4$, $p = 0.5$

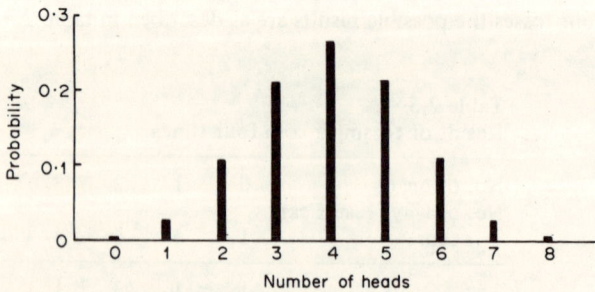

Fig. 2.3 Probability distribution for tossing a coin eight times; $n = 8$, $p = 0.5$

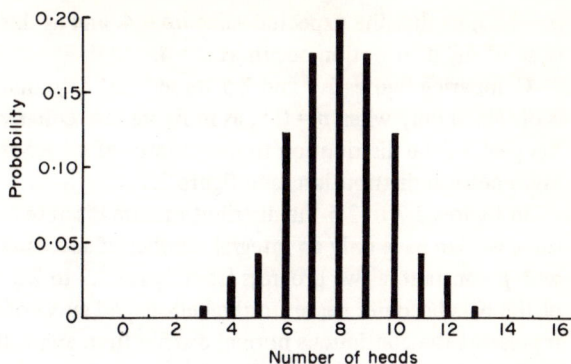

Fig. 2.4 Binomial probability distribution for tossing a coin sixteen times; $n = 16$, $p = 0.5$

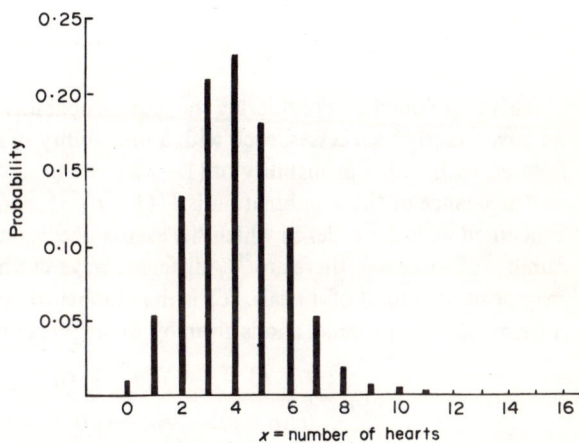

Fig. 2.5 Asymmetrical binomial distribution. Probability of drawing x hearts in sixteen attempts from a pack of playing cards; $n = 16$, $p = 0.25$

in a single event $p = 1 - p = 0.5$. This is the probability of, for example, tossing a head with an unbiased coin.

If we select a different probability, for example, $p = 0.25$, corresponding to the probability of obtaining, say, a heart in any particular draw from a pack of playing cards, we obtain an asymmetrical distribution. For example, the probability distribution for the total number of hearts drawn in sixteen attempts is shown in figure 2.5. Here $n = 16$,

$p = 0.25$, so that the expected value $np = 4$, and as this is an integer the peak of the distribution occurs at $x = 4$.

Comparing figures 2.4 and 2.5 we see that a symmetrical distribution is obtained only when $p = 0.5$, as in figure 2.4. Other values of p shift the peak of the distribution to the vicinity of $x = np$ and thus give an asymmetrical distribution as in figure 2.5.

In figures 2.2 to 2.5 the distributions are discrete or discontinuous, since we can have only an integral number of successes. As n increases with p constant, as we progress from figure 2.2 to 2.5, the envelope of the bars becomes nearer to the bell-shaped curve of figure 2.9. This represents the continuous normal distribution, and is the limiting case of the binominal distribution as n becomes very large (section 2.21).

A more general case is that of an event with a probability of success of p. If we make n independent trials, the probability of r successes in n trials is given by

$$P(r) = {}^{n}C_r \times P^r \times (1 - P)^{n-r} \qquad (2.12)$$

This is obtained by considering the joint probability that in n events we have exactly r successes, each with a probability of P, and $n - r$ failures, each with a probability of $(1 - p)$.

The chance of this combination is $P^r(1 - p)^{n-r}$, and as we are not concerned with the order in which the events occur, only with the total number of successes, there are ${}^{n}C_r$ different ways in which r successes may occur in a total of n trials. ${}^{n}C_r$ is the number of combinations of n items taken r at a time, and is given by the expression

$$ {}^{n}C_r = \frac{n!}{r!(n-r)!} = \frac{n(n-1)(n-2)(\ldots)(n-r+1)}{r(r-1)(r-2)(\ldots)(\times 3 \times 2 \times 1)} $$

Since each of the ${}^{n}C_r$ ways in which r successes can occur is equally probable, the total probability of r successes is ${}^{n}C_r$ times the probability $p^r(1 - p)^{n-r}$ of r specified trials being successful and the remaining $(n - r)$ being failures.

Hence the total probability of r successes is

$$P(r) = {}^{n}C_r \times P^r(1 - P)^{n-r}$$

The above expression for $P(r)$ enables us to assign a probability to each value of r from 0 to n. The resulting distribution of probabilities is known as the binomial distribution since the coefficients such as ${}^{n}C_r$ are the coefficients of x^r in the binomial expansion of $(1 + x)^n$ as a power series. An alternative name is the Bernoulli distribution, after the mathematician who first derived it.

The binomial distribution can be identified more closely with the binomial series if we consider the expansion of $[p \times x + (1-p)]^n$.

The term containing x^r in this expansion is

$$U_r = {}^nC_r \times p^r \times x^r \times (1-p)^{n-r}$$
$$= x^r \times p(r)$$

showing that the probability $p(r)$ of r successes is equal to the coefficient of x^r in the expansion. If now we make $x = 1$, we have

$$[p + (1-p)]^n = 1$$
$$= p(0) + p(1) + p(2) + \ldots + p(r)$$
$$+ \ldots + p(n)$$

Thus the sum of the probabilities of obtaining $0, 1, 2, 3 \ldots n$ success is 1, as expected, since there are no other possible outcomes.

2.12   Examples of Binomial Distribution

For an example of the general case, as given in equation (2.12), let us take an electronic system which has a reliability of 0.95 for a single day's operation. What is the probability of experiencing a failure on two days in a forty day period?

In this example the probability of a fault on any one day is $1 - 0.95 = 0.05$, since the reliability is the probability of not experiencing a fault. Thus $p = 0.05$ and $(1 - p) = 0.95$.

The probability we require is then

$$p(2) = {}^{40}C_2 \times p^2 \times (1-p)^{38}$$
$$= \frac{40 \times 39}{2 \times 1} \times \left(\frac{1}{20}\right)^2 \times 0.95^{38}$$
$$= \frac{40 \times 39}{2 \times 400} \times 0.95^{38}$$
$$= \frac{39}{20} \times 0.1425 = \underline{0.2778}$$

The probability of 0, 1, 3, etc. faults during the forty day period can be found in a similar manner, the results being shown in table 2.4.

Table 2.4

No. of days with fault	Probability
0	0·1285
1	0·2705
2	0·2778
3	0·1851
4	0·09012
5	0·03415

Since the probability of a fault on any particular day is 0·05, the expected number of days with a fault in a forty day period is 40 × 0·05 = 2 days. This is the outcome which is most probable from the figures of table 2.4.

The binomial distribution, although applicable to many practical cases, is valid only in certain conditions, some of which have been mentioned above. For reference, they are collected together in the following list

1 The probability of success in one trial is unaffected by the success or failure of other trials; that is, the trials are independent.
2 Each of the trials has only two possible results, called for convenience 'success' and 'failure'.
3 The probability of success is the same for each trial, since success and failure are complementary events; if the probability of success is constant and equal to P, the probability of failure is equal to $(1 - P)$ and is also constant.

The third condition generally includes the first condition, for example, in drawing cards at random from a pack. If each card drawn is held and not returned to the pack for the next draw, the probability of drawing, say, an ace is different at each trial, since each draw takes place from a different number of cards. Also the probability of drawing an ace depends upon the number of aces previously drawn and thus removed from the pack.

In this instance both the first and third conditions are not satisfied. If, however, we return each card drawn to the pack before the next draw, we are drawing a card each time from the same set of fifty-two cards with the same probability of selecting an ace at each trial. In this case both the first and the third conditions are satisfied. The second condition is also satisfied since we can consider drawing an ace as success and drawing any other card as a failure. Thus the binomial distribution is directly applicable under these modified conditions.

The previous example of the application of the binomial theorem was concerned with the calculation of the probability for a particular number of faults. Where we have some spare equipment, we may require the sum of the probabilities of several different outcomes. As an example, we examine the reliability of a small radio transmitter used intermittently in which the major source of failure is the thermionic valves. This equipment has nine similar working valves and one spare, which can be substituted for a faulty valve in a few minutes without materially interrupting the service provided. If each valve has a reliability of 0·95 during the period between maintenance checks and the replenishment of spares, what is the reliability of the transmitter during this interval?

Neglecting failures in any other components, the transmitter will be operational if at the end of the interval between routine maintenance checks, either nine or all ten of the valves are working. The probability of all ten working is, by the product rule

$$P_{10} = (0·95)^{10}$$

since the reliability $p = 0·95$ is the probability that any particular valve is operational. From the binomial distribution, the probability that nine of the valves are working and one is faulty is

$$P_9 = 10 \times (0·95)^9 (1 - 0·95)^1$$
$$= \underline{0·5(0·95)^9}$$

The combined probability that either nine or ten are working is found by adding P_9 and P_{10}, since these results are mutually exclusive.

Thus the required reliability R is given by

$$R = P_{10} + P_9$$
$$= (0·95)^{10} + (0·5)(0·95)^9$$
$$= (0·95 + 0·5)(0·95)^9 = 1·45(0·95)^9$$
$$= \underline{0·914}$$

For a transmitter without a spare valve, the reliability over the same period is

$$R_1 = (0·95)^9$$
$$= \underline{0·630}$$

since all nine valves must remain working.

Thus the provision of an extra valve raises the reliability from 0·630 to 0·914 over the same period. The probability of failure is reduced from

$$1 - 0·630 = \underline{0·370}$$

to

$$1 - 0·914 = \underline{0·086}$$

during the same period of operation. This is a reduction by a factor of $0·370/0·086 = \underline{4·3}$.

2.13   The Most Probable Outcome

One interesting characteristic of the binomial distribution is the number of successes most likely to occur. In the example on p. 40, the probabilities listed in table 2.4 show that the most probable number of successful events is two.

$$\text{Here } n = 40, p = 0·05, \text{ thus } np = 2.$$

According to the relative frequency view of probability, if the series of forty trials were repeated many times, the average number of successful events would be two in each set of forty trials. Thus the expected number of successes is two, and as this is an integer, the most likely number of successes as found from the binomial distribution is also two.

In general, the expected number of successes in the series, np, will not be an integer. Since we must have an integral number of successes, we cannot say that k, the most likely number of successes, is equal to np. We can, however, say that k will be equal to one of the integers on either side of np.

Thus, if the data is altered slightly so that the probability of a fault on any day is 0·06, the expected number of days on which a fault occurs in a forty day period is

$$np = 40 \times 0·06 = 2·4 \text{ days}$$

Using the binomial distribution the probability of faults occurring on two days of the forty is

$$p(2) = \frac{40 \times 39}{1·2} \times (0·94)^{38}(0·06)^2$$

$$= \underline{0·2675}$$

The probability of faults occurring on three days out of the forty is

$$p(3) = \frac{40 \times 39 \times 38}{1 \times 2 \times 3} \times (0 \cdot 94)^{37}(0 \cdot 06)^{3}$$

$$= \underline{0 \cdot 2162}$$

Finally, the probability of a fault occurring on only one day is

$$p(1) = 40 \times (0 \cdot 94)^{39}(0 \cdot 06)$$

$$= \underline{0 \cdot 2149}$$

Thus the most likely outcome is two faults during the period, the chance of either one or three faults being considerably smaller. The expected number of faults is $2 \cdot 4$, since $0 \cdot 06$ is the probability of a fault on any day, and the complementary probability of no fault is $0 \cdot 94$. Thus, the most likely number of faults is in this case the next integer below $np = 2 \cdot 4$.

2.14 The Poisson Distribution

It is evident from this example, and from the diagram in figure 2.6, that the expected number of successes, np, decides where the peak of the probability distribution occurs. It thus appears that the product np is the most important characteristic of the binomial distribution. However, in order to evaluate the individual terms of the distribution we need to know n and p separately. In some circumstances n is not easily defined, or may not be known, whereas the product np is known. We are thus led to enquire whether any deductions can be made from a knowledge of np alone, when neither n nor p is known separately.

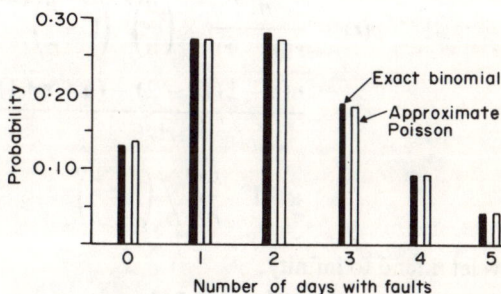

Fig. 2.6 Comparison of binomial distribution and Poisson approximation for $n = 40$, $p = 0 \cdot 05$

Although this may seem a meagre amount of information, there is in fact a probability distribution which requires this fact alone, and which is applicable in very many practical situations. It is called the Poisson distribution, after its originator, and it can be obtained directly from the binomial distribution if np is held constant as n tends to infinity. The Poisson distribution is thus the limiting case of the binomial distribution as n, the number of trials, becomes very large, and the probability of success per trial, p, becomes small.

If we consider that success denotes the failure of a component which is one of a large number on test, the reliability of modern components ensures that failure is a comparatively rare event, and the Poisson conditions are satisfied. As with the binomial distribution the probability of failure per trial or per unit time must be constant, and it must not depend upon the success or failure of any other trial.

The Poisson distribution is applied to a variety of situations in which events occur randomly but the above conditions are satisfied; history relates that Poisson first applied it to the statistics of the annual number of cavalrymen in Napoleon's army who were kicked to death by horses!

Although the derivation of the Poisson distribution as the limiting case of the binomial theorem is not the only method, it is probably the simplest. We start with the rth term of the binomial distribution, corresponding to r successes out of n trials

$$p(r) = \frac{n!}{r!(n-r)!} \times p^r \times (1-p)^{n-r}$$

We now make μ the expected number of successes by setting $\mu = n \times p$.

Thus

$$p(r) = \frac{n!}{r!(n-r)!} \times \left(\frac{\mu}{n}\right)^r \left(1 - \frac{\mu}{n}\right)^{n-r}$$

$$= \frac{n(n-1)(n-2) - (n-r+1)}{r!} \times \frac{\mu^r}{n^r}$$

$$\times \left(1 - \frac{\mu}{n}\right)^n \Big/ \left(1 - \frac{\mu}{n}\right)^r$$

We now let n tend to infinity.

Since n is very much greater than r

$$n(n-1)(n-2) \ldots (n-r+1) \text{ tends to } n^r$$

Also since there are only a finite number of factors, each tending to 1

$$\left(1 - \frac{\mu}{n}\right)^r \quad \text{tends to 1}$$

Thus

$$\lim_{n \to \infty} [p(r)] = \frac{n^r}{r!} \times \frac{\mu^r}{n^r} \times \lim_{n \to \infty} \left(1 - \frac{\mu}{n}\right)^n \Big/ 1$$

$$= \frac{\mu^r}{r!} \times \lim_{n \to \infty} \left(1 - \frac{\mu}{n}\right)^n$$

$$= \frac{\mu^r}{r!} \times \exp(-\mu) \qquad (2.13)$$

Since

$$\lim_{n \to \infty} \left(1 + \frac{1}{n}\right)^n =$$

and

$$\lim_{n \to \infty} \left(1 + \frac{x}{n}\right)^n = \exp(x)$$

this expression gives the probability of r successes occurring when the expected number is μ. By putting $r = 0, 1, 2, 3$, etc., we obtain the Poisson series which shows the probability of obtaining $0, 1, 2, 3$, etc., successes. The series is

$$S = \exp(-\mu) + \mu \exp(-\mu) + \frac{\mu^2}{2!} \exp(-\mu) + \frac{\mu^3}{3!} \exp(-\mu)$$

$$+ \dots + \frac{\mu^r}{r!} \exp(-\mu) + \dots \qquad (2.14)$$

This series may be written as

$$S = \exp(-\mu) \left(1 + \mu + \frac{\mu^2}{2} + \frac{\mu^3}{3} + \dots + \frac{\mu^r}{r} + \dots\right)$$

The sum to infinity of the series in brackets is $\exp(\mu)$, so that the sum to infinity of the series S is

$$S = \exp(-\mu) \times \exp(\mu) = \exp(0) = 1$$

Thus, as expected, if we add together the probabilities of 0, 1, 2, 3, etc., successes, we include all possible outcomes, and the total probability must be 1.

The values of the individual terms of the series S give the following

$$\exp(-\mu) \qquad = \text{probability of zero successes}$$

$$\mu \exp(-\mu) \qquad = \text{probability of one success}$$

$$\frac{\mu^2}{2} \exp(-\mu) \qquad = \text{probability of two successes}$$

and

$$\frac{\mu^r}{r!} \times \exp(-\mu) = \text{probability of } r \text{ successes}$$

If success means the occurrence of a fault in a system or component, the case of zero successes is that of zero faults. This is the probability which is important in studying the reliability of non-redundant systems, that is, systems in which no spares are available in the event of a failure.

Where spares are available and can be used to maintain or restore service, further terms in series may be required to assess reliability.

2.15　Examples of Poisson Distribution

As an example of the use of the Poisson distribution, we consider the reliability of a batch of mobile radio transmitters. Used in the same circumstances, these are found to experience on average one fault in each set every eighteen weeks. We require the probability that any one set will operate for (a) 5 weeks, (b) 20 weeks without a fault.

The first quantity to be calculated in the expected number of faults. In general terms this is given by

$$\mu = \frac{\text{period of operation}}{\text{mean time between faults}}$$

This is 5/18 for case (a), 20/18 for case (b).

Thus, for the 5 week period the required probability is given by the first term of expression (2.14), that is

$$p(5) = \exp(-5/18) = \exp(-0.278) = \underline{0.762}$$

Similarly, for the 20 week period, the probability of no faults is

$$p(20) = \exp(-20/18) = \exp(-1 \cdot 11) = \underline{0 \cdot 329}$$

Clearly, the computation involved in using the Poisson distribution is much less than that required to evaluate the binomial distribution for large values of n. But we have seen that the Poisson formulae are the exact equivalents of the binomial formulae only when n tends to infinity. However, for many practical situations an approximate result is adequate, since the basic data are generally of only limited accuracy. We could thus save much time when calculating probabilities if we could use the Poisson distribution as an approximation to the binomial distribution for large values of n, and thus avoid calculating quantities like $(0 \cdot 94)^{38}$ required in a previous example.

For most purposes the Poisson distribution is a reasonable approximation to the binomial distribution if $n > 20$ and $p < 0 \cdot 05$.

As an example of the accuracy obtainable, we return to the case of an electronic system which has a probability of failure on any one day of $0 \cdot 05$, and we calculate the probability of various numbers of failures in a forty day period.

The expected number of failures is

$$\mu = np = 40 \times 0 \cdot 05 = 2 \text{ per period}$$

Thus, the probability of 0 failures is

$$p(0) = \exp(-\mu) = \exp(-2) = 0 \cdot 1353$$

The probability of 1 failure is

$$p(1) = \mu \exp(-\mu) = 2 \exp(-2) = 0 \cdot 2706$$

The probability of 2 failures is

$$p(2) = \frac{\mu^2}{2} \exp(-\mu) = 2 \exp(-2) = 0 \cdot 2706$$

Further probabilities may be calculated in a similar manner. The value of $p(1)$ is very near to the exact value given by the binomial theorem. The two distributions are compared in table 2.5 below, and show for these values of p and n fairly good agreement.

The major discrepancy in this example is the value of $p(2)$. The binomial distribution shows that, since the expected number of faults is 2, $p(2)$ is the greatest probability, whereas the Poisson distribution gives $p(2) = p(1)$.

Table 2.5

No. of days with fault	Exact probability (binomial)	Approximate probability (Poisson)
0	0·1285	0·1353
1	0·2705	0·2706
2	0·2778	0·2706
3	0·1851	0·1804
4	0·09012	0·0902
5	0·03415	0·0361

2.16   The Exponential Failure Law

A very useful form of the Poisson probability distribution can be obtained for systems with a constant failure rate by expressing the expected number of events, μ in terms of the operating period of some component or system, and its mean time between failures.

If the MTBF is M, we expect one fault in a period M, and thus T/M in a period T. Thus $\mu = T/M$, and the Poisson distribution of equation (2.14) becomes

$$S = \exp(-T/M) + \frac{T}{M}\exp(-T/M) + \tfrac{1}{2} \times \frac{T}{M}^{2}$$
$$\times \exp(-T/M) + \text{etc.} \tag{2.15}$$

Whence the probability of zero failures is

$$p(0) = \exp(-T/M)$$

This expression is important for computing reliability, and is shown graphically in figures 2.7 and 2.8.

The probability of one failure is

$$p(1) = \frac{T}{M}\exp(-T/M)$$

The probability of r failures is

$$p(r) = \left(\frac{T}{M}\right)^{r} \times \frac{1}{r!} \times \exp(-T/M) \tag{2.16}$$

For reliability calculations on systems which have no redundancy, the first term is the important one; this gives the probability of zero failures during the specified operating period, that is, the reliability.

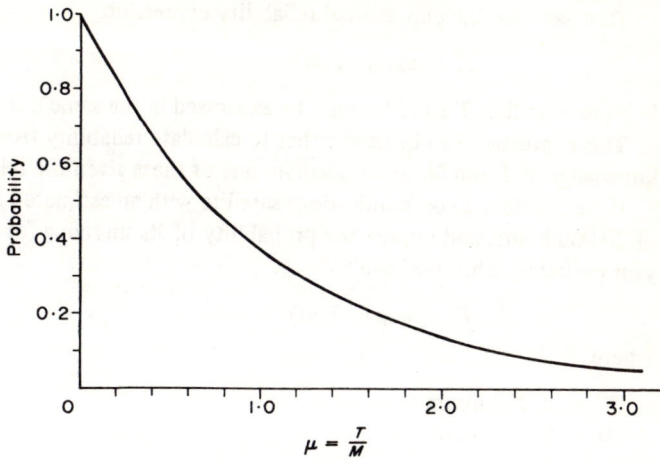

Fig. 2.7 Graph of exponential failure law on linear scale

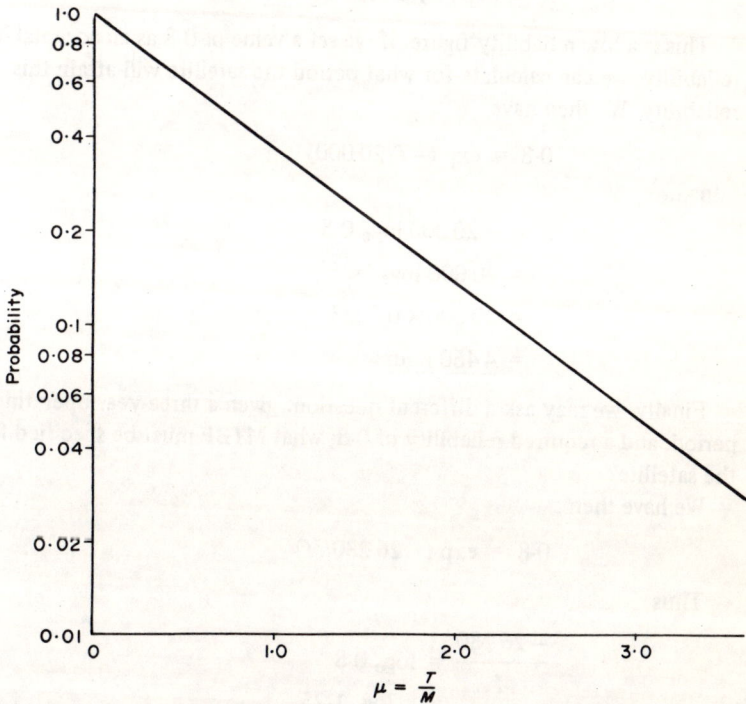

Fig. 2.8 Graph of exponential failure law on logarithmic scale

Thus we have the exponential reliability expression

$$R = \exp(-T/M) \tag{2.17}$$

It is essential that T and M should be expressed in the same units.

This expression can be used either to calculate reliability from a knowledge of T and M, or to calculate one of these given the reliability.

If we consider a communications satellite with an estimated MTBF of 20 000 hours, and require the probability of its surviving for a three-year period in orbit, we have

$$R = \exp(-T/M)$$

where

T = 26 280 hours
M = 20 000 hours

then

$$R = \exp(-1 \cdot 314) = \underline{0 \cdot 268}$$

This is a low reliability figure. If we set a value of 0·8 as an acceptable reliability, we can calculate for what period the satellite will attain this reliability. We then have

$$0 \cdot 8 = \exp(-T/20\,000)$$

Thus

$$T = -20\,000 \log_e 0 \cdot 8$$
$$= 20\,000 \log_e 1 \cdot 25$$
$$= 20\,000 \times 0 \cdot 2231$$
$$= \underline{4\,460 \text{ hours}}$$

Finally, we may ask a different question: given a three-year operating period, and a required reliability of 0·8, what MTBF must be specified for the satellite?

We have then

$$0 \cdot 8 = \exp(-26\,280/M)$$

Thus

$$\frac{-26\,280}{M} = \log_e 0 \cdot 8$$
$$= -\log_e 1 \cdot 25$$
$$= -0 \cdot 2231$$

Whence

$$M = \frac{26\,280}{0 \cdot 2231}$$

$$= \underline{117\,800 \text{ hours}}$$

This is about 13½ years.

All of the above calculations are founded on the assumption that the system cannot be repaired if a fault occurs, and that it contains no spare units which could maintain service in the event of a failure. These assumptions are not necessarily true for all systems, and in such circumstances other methods of assessing system reliability must be used.

2.17 Statistics

In addition to a knowledge of elementary probability theory, the reliability engineer requires more knowledge of the principles of statistics, and of their application to his problems. Statistics has been described as the collection and arrangement of data concerning a particular group of individuals, in order to extract the salient features from a large accumulation of facts. The individuals in the group must have some identifiable property which distinguishes them from other persons or objects. We may thus collect data on peoples' ages at death for mortality tables, or on the life of various types of electronic component for reliability predictions.

In all of these cases we are faced with a mass of data, and we use statistical methods in order to extract from it a few numerical values which convey the important characteristics of the data. This process of extraction clearly involves the rejection of a considerable amount of detailed information, but the advantage of obtaining a few numerical measures of the particular population rather than a mass of data outweighs the loss. Only by such a condensation is it possible to reduce the data to a form suitable for calculation. For example, many authorities are concerned with the numbers and ages of the country's population, for predicting the need for schools or houses, or for life insurance. Clearly, the raw data consisting of the age of every member of the population (over fifty million items) would be entirely unmanageable and some data reduction process is essential.

2.18   The Mean

The most useful single fact about a collection of data is a numerical
estimate of its general magnitude, usually called a measure of location.
There are several of these, the most widely used being the average, or
more exactly the arithmetic mean. This is obtained by adding together
all of the magnitudes of the items, and dividing by the number of items.
Thus, if the magnitude of the rth item is x_r, the mean of the total of
n items is

$$M = \frac{1}{n} \sum_{1}^{n} x_r \tag{2.18}$$

For example, if a batch of items are tested and the times to failure are
recorded as 422, 403, 511, 451, 462 hours, the mean time to first
failure is given by

$$M = \frac{\sum x_r}{n} = \frac{422 + 403 + 511 + 451 + 462}{5}$$

$$= \frac{2\,249}{5} = \underline{449{\cdot}8 \text{ hours}}$$

Since the basic data is rounded off to the nearest hour, the MTFF is
best rounded in the same way to $\underline{M = 450 \text{ hours.}}$

2.19   Other Measures of Location

Two other measures of location which are useful are the mode and the
median. The mode is the value which occurs most frequently. It will
differ from the arithmetic mean unless the distribution of values is
symmetrical above and below the mean. The mode is a useful concept
in most distributions of data where there is a central peak, corresponding
to the value of the variable which occurs most frequently. However,
more complex situations may occur in which there are several peaks,
and in this case there is no unique value of mode. For example, the
following data relate to the numbers of marks obtained by some
candidates for an examination

Mark	7	8	9	10	11	12	13	14
No. of students	11	16	19	17	16	10	5	3

From this the maximum frequency is nineteen, and the corresponding mark is nine. Thus the value of the mode is nine marks.

The median is the central value in a set of data. It is found by arranging the data in ascending order of magnitude and then choosing the value which has as many items above it as below it.

2.20　Variability

Although the mean is generally the most useful single quantity we can extract from a mass of data, we have obviously rejected much information by reducing our data to a single number. This loss is most apparent when we are enquiring into the variability of the items of data. The mean tells us the general magnitude of the items but nothing about the way the magnitudes are scattered above and below the mean. For this purpose we require a different measure which indicates dispersion or variability.

A number of different measures of dispersion have been proposed, the simplest of which is the range. This is merely the difference between the maximum and minimum values of the variable. In the above table of marks the range is

$$R = 14 - 7 = 7 \text{ marks}$$

Although this is simple to compute, it is not very satisfactory as it depends upon only two items in the set of data. A much better measure of dispersion is obtained by using all of the data given. This is most conveniently effected by considering divergences from the mean value. Thus, if M is the arithmetic mean of a set of data $x_1, x_2, x_3 \ldots x_n$ the deviation of x_r is $(x_r - M)$, and a possible measure of variability is the average value of $|x_r - M|$. (The absolute value is taken since we are not concerned with the sign of the deviation.)

However, a more useful measure is the root mean square deviation. This is obtained by squaring each value of $(x_r - M)$, averaging the squares, and then extracting the square root. Using the above symbols, this is

$$S = \sqrt{\left(\frac{1}{n} \times \sum_{r=1}^{n} (x_r - M)^2 \right)} \tag{2.19}$$

This is called the standard deviation, and its square is called the variance.

For example, in the times to failure given in section 2.18 the five deviations $(x_r - m)$ are

$$450 - 422 = 28 \text{ hours}$$
$$450 - 403 = 47 \text{ hours}$$
$$511 - 450 = 61 \text{ hours}$$
$$451 - 450 = 1 \text{ hour}$$
$$462 - 450 = 12 \text{ hours}$$

The mean square value of the deviations is given by

$$(28^2 + 47^2 + 61^2 + 1^2 + 12^2) \times \frac{1}{5} = 1372$$

and the standard deviation is the square root of this, namely

$$S = 1372 = 37 \text{ hours}$$

to the nearest hour.

For large sets of data, it may be easier to calculate merely the sums of the squares of the items, and use the relation

$$S^2 = \frac{1}{n} \sum_{r=1}^{n} x_r^2 - M^2 \qquad (2.20)$$

where M is the mean value of the items $x_1, x_2 \ldots x_n$.

The formula for S^2 is particularly useful if the values of x are being read into a computer, since it avoids the need to store all values of x. As the data is fed in, the sums of the items and of the squares of the items are accumulated. At the end of the read-in phase, the mean value M can be calculated, and then the above expression gives S^2.

2.21 The Normal Distribution

A further distribution is of interest in considering probability, the 'normal' or Gaussian distribution. This is particularly important in considering the variations in a number of manufactured articles, which should all be identical if the process were carried out under ideal conditions with perfect machinery. In practical conditions we cannot maintain all conditions constant indefinitely; there are bound to be small variations in the characteristics of the raw material, in the performance of all tools used in the process due to wear, backlash, etc., and in the ambient conditions. Thus, we find that if we attempt to machine a set of nominally identical components in an automatic lathe, there will be small variations between them due to a variety of causes. A similar result occurs in the production of electrical components. If we attempt to make a batch of 100 ohm resistors we find that their values will not all be equal, but if the process is well controlled they will all cluster near to the desired value of 100 ohms.

In both of these examples we find that a plot of component value against the frequency of its occurrence approximates to a bell-shaped curve called the normal curve, or the normal error curve. There are only two quantities needed to define a normal curve: the mean value and the standard deviation. For many purposes it is convenient to rescale the variable so that the average value is zero. If the deviation of the variable from the mean is then expressed in terms of the standard deviation, we obtain a single universal curve which fits all normal probability distributions.

The mathematical expression for the normal error curve is

$$y = \frac{1}{\sigma\sqrt{(2\pi)}} \times \exp\left[\frac{-(x-\mu)^2}{2\sigma^2}\right] \qquad (2.21)$$

where y = ordinate of curve

x = variable

σ = standard deviation of variable

μ = mean value of variable

The ordinate for any value of x is proportional to the probability of occurrence of that value of x in the population.

If this general expression is rescaled so that the mean value is zero and the deviation is scaled in units of σ, that is, a new variable t is used, related to x by the expression

$$t = \frac{x-\mu}{\sigma}$$

we have

$$y = \frac{1}{\sigma\sqrt{(2\pi)}} \times \exp(-t^2/2)$$

If the unit in this is taken as one standard deviation, we get the universal curve

$$y = \frac{1}{\sqrt{(2\pi)}} \times \exp(-t^2/2) \qquad (2.22)$$

The shape of this curve is illustrated in figure 2.9.

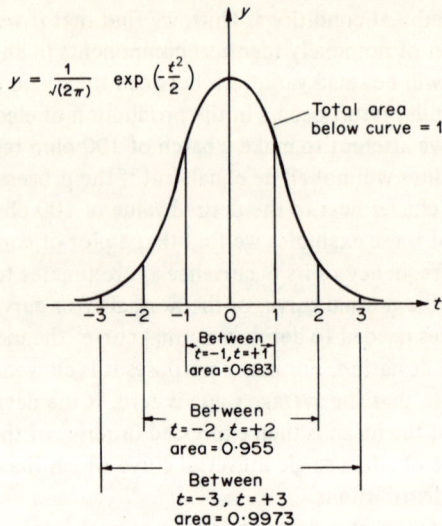

$$y = \frac{1}{\sqrt{(2\pi)}} \exp\left(-\frac{t^2}{2}\right)$$

Total area
below curve = 1

Between
$t = -1, t = +1$
area = 0.683

Between
$t = -2, t = +2$
area = 0.955

Between
$t = -3, t = +3$
area = 0.9973

Fig. 2.9 Normal error curve

This curve is a continuous one, of the kind found by measuring a set of items when each item can have any magnitude. Sometimes it is convenient to consider classes of magnitude and count the number of items in each class. Thus, we may have an automatic grading machine which measures a batch of nominal 100 ohm resistors and classifies them into 90 to 90·9 ohms, 91 to 91·9 ohms, 92 to 92·9 ohms, etc. The results of such a measurement would be a discontinuous or discrete distribution which could be represented by a histogram. The shape of the histogram would approximate to a normal error curve, becoming nearer to it as the number of classes were increased.

An important feature of the normal error curve is the area between any two ordinates. This gives the proportion of the total population having magnitudes between those of the corresponding abscissae. Thus, using the curve of figure 2.9 we see that between $t = +1$ and $t = -1$ lies an area of the curve equal to 0·683. Thus, if we have a set of nominal 100 ohm resistors whose values are normally distributed, with mean equal to 100 ohms and standard deviation equal to 10 ohms, 68·3 per cent of them will on average lie between 90 and 110 ohms ($t = -1$ and $t = +1$) and 95·4 per cent between 80 and 120 ohms ($t = -2$ and $t = +2$).

Alternatively, we can say that if we take any resistor at random, there is a probability of 0·683 that it will lie between 90 and 110 ohms, 0·954 that it will lie between 80 and 120 ohms, etc.

Unfortunately the mathematical operation of integration, required to find the area below a curve, cannot be performed directly on the normal error curve, since there is no function which, when differentiated, has the form $y = \exp(-t^2)$. Consequently we can obtain information about the area below the curve only from tables which have been constructed by numerical integration. These are widely used and appear in most books of mathematical tables.

The normal distribution can be regarded as the limiting case of a binomial distribution when n, the number of trials, increases indefinitely, and it is often simpler to use the normal error function tables for cases of the binomial distribution where n is large. In equation 2.12 the approximation is a good one for $p \leqslant 0·5$ and $np \geqslant 10$, or $p > 0·5$ and $n(1 - p) > 10$.

An interesting feature of the normal distribution is that if we consider a variable x which is the sum of a number of random distributions, the variable x will have approximately a normal distribution, even though the individual distributions which make it up may have different distributions. This is the reason for so many sets of data having a normal distribution; it matters little how the variations due to each cause are distributed, so long as many causes of variability are present.

Experiments with random numbers on a digital computer suggest that about six causes are sufficient to give a good approximation to normal distribution. If we start with six sets of random numbers evenly distributed between the same limits, and form a new distribution by adding together a number from each set, the new distribution is a good fit to the normal error curve, and is acceptable as a normal distribution for many engineering applications. More accuracy can be obtained at the price of lower speed by taking more than six sets of numbers.

This chapter contains most of the mathematics required to tackle the basic problems in reliability engineering, and the results given will be used to analyse a variety of systems in the later chapters.

PROBLEMS

1. A tray contains sixteen resistors, five within tolerance, seven below tolerance and four above tolerance. Three resistors are selected at random. What is the probability that they are all within tolerance? [1/56]

2. An unattended radio transmitter contains five identical units operating in parallel, any four being sufficient to provide the required aerial drive. If the probability of failure of any single unit between maintenance checks is 0·10, what is the probability that the transmitter will not fail during this period?   [0·919]

3. A laboratory buys sixty oscilloscopes from supplier A, twenty-five from B and fifteen from C. The probability of any one being faulty is 0·05 for A, 0·10 for B, and 0·15 for C. What is the probability that any one selected at random from the 100 will work correctly? [0·9225]

4. A particular type of transistor is tested for three defects, low gain (A), high collector leakage (B) and low collector voltage breakdown (C), which may occur singly or in combination. The probability of A is 0·5 per cent, of B is 0·4 per cent, and of C is 0·4 per cent. The probability of both A and C is 0·12 per cent, of both B and C is 0·1 per cent, and of both B and C is 0·08 per cent. The probability of all three defects together is 0·02 per cent. What is the probability that any particular transistor has at least one of the three faults?   [1·02 per cent]

5. The probability of a power failure in a period of one year at a particular location is 6 per cent. What is the probability of experiencing (a) no failure, (b) at least one failure during a five-year period? [(a) 74·1 per cent; (b) 25·9 per cent]

6. A car battery manufacturer finds that the lives of a popular type of battery have a normal distribution with a mean of thirty months and a standard deviation of 5·2 months. What percentage of batteries must he expect to exchange under guarantee, if this extends for two years? (This question requires tables of the normal error function.) [12·4 per cent]

7. Records of the lives of incandescent lamps show the following results: 943, 886, 1 057, 901, 1 113, 1 094, 873, 917, 1 151, 892, 1 121, 1 186 hours. Calculate the mean and standard deviation of the sample. If this sample is representative of a batch of 1 000 lamps, and their lives are normally distributed, how many of them are likely to have lives less than 898 hours? (Use the data of figure 2.9.)
[(a) Mean = 1 011 h; (b) Standard deviation = 113 h; (c) About 159 lamps]

3 Reliability Prediction

3.1 System Subdivision

In this chapter we investigate the reliability of complete systems in terms of the reliabilities or the failure rates of the components or units used to construct them. The first requirement is thus to divide the system into units, each small enough to enable its reliability to be assessed directly. In general this requires subdivision down to component level so that we may use reliability data gathered from many different systems which incorporate the same components. Although we regard as components individual items such as a resistor, a transistor or a capacitor, the introduction of integrated circuits of increasing complexity has involved some alteration to the definition of a component.

If we define a component as an item which cannot be subdivided without destroying it, or as the smallest item which could be replaced when repairing a fault, we must classify an integrated circuit, however complex as a component. Even though it may embody many transistors, resistors and diodes these elements are not individually accessible, and the circuit can be used and tested only as a complete entity.

In addition to information about the number and characteristics of all the components in a system we require knowledge of the way in which they are connected together. This provides the essential data for reliability assessment.

3.2 Reliability Models

A useful analytical tool for predicting system behaviour is the reliability model or reliability diagram. This illustrates the functional relation between the various components of the system and the way in which a failure of each component would affect the overall system performance.

The simplest arrangement is one in which each component is necessary for correct operation of the system, and in consequence a failure of any one component must cause a system failure.

This is often called a 'series system' since we may regard each component working properly as a closed switch. If all of the components are connected in series, we obtain a conducting path through the network only when all elements are conducting, corresponding to the condition of the system where all components are functioning correctly. The two diagrams are shown in figure 3.1.

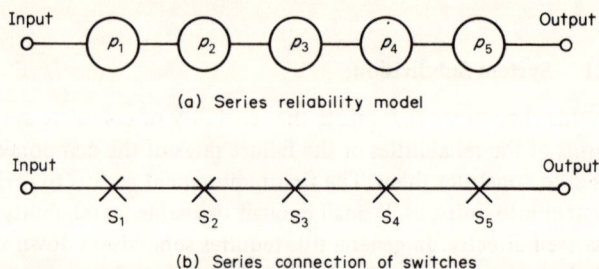

(a) Series reliability model

(b) Series connection of switches

Fig. 3.1 Series reliability system

If we examine this system to determine overall reliability directly, we require the overall probability of an interval with zero failures in all of the five separate components. Thus the condition of zero failures must obtain simultaneously in all components. This is the problem of joint probability discussed in section 2.5 in which the probability of all five events occurring is equal to the product of the separate probabilities of the event occurring in each component. The 'event' in this case is the occurrence of zero failures in some predetermined interval.

Thus, if the separate reliabilities of the five components (the probabilities of zero failures) are $R_1, R_2, R_3, R_4,$ and R_5, the overall probability of no system failure, that is, the overall system reliability, is

$$R = R_1 \times R_2 \times R_3 \times R_4 \times R_5 \qquad (3.1)$$

This is fairly simple to evaluate in this example, but may be tedious if there are a large number of components, each having a reliability very close to 1. Many significant figures must then be retained in the calculation in order to obtain adequate overall accuracy. This approach is thus not convenient for computation and an alternative approach may be simpler and more accurate. This involves essentially transferring our attention to the small amount by which each component reliability figure falls below 1, or in other words, examining the probability of failure rather than success.

3.3   Failure Analysis of Series System

In this method of analysis we can use the same reliability diagram but examine the conditions for failure. Since we are dealing with a series system a failure in any component will cause a system failure. Thus the probability we now require is that of a set of alternative situations, the probability that either component 1 or component 2 or component 3, etc., will develop a fault during a prescribed interval. This is the situation discussed in section 2.7. If the separate probabilities are mutually exclusive, the joint probability is found by the addition rule, so that if the separate probabilities of a fault developing in the five components are p_1, p_2, p_3, p_4, p_5, the joint probability of a fault occurring in one or more of the five components is

$$P = p_1 + p_2 + p_3 + p_4 + p_5 \qquad (3.2)$$

In practice, of course, the various events are not mutually exclusive since it is possible for faults in components 1 and 2 to occur together. However, in a system having any useful degree of reliability the joint probability is much smaller than the probability of a single fault occurring. Thus, the error involved in neglecting the probability of multiple faults is not significant.

Equation (3.2) is not the most convenient form for general use, since the probabilities p_1 to p_5 depend upon the duration of the test or the prescribed operating period. If this period is changed the probabilities of failure will change. A more useful formula for calculating reliability would be one involving time explicitly, rather than indirectly. This can be developed by using a characteristic of the components which is generally independent of time — the failure rate. This gives the expected number of failures per unit time. If the failure rates for the components are $\lambda_1, \lambda_2, \lambda_3$, etc., the expected number of failures are

$$n_1 = \lambda_1 T \text{ for component 1}$$
$$n_2 = \lambda_2 T \text{ for component 2, etc.}$$

where T is the duration of the test, or operating period. Since we are considering a series system, any one of these faults will cause a system failure. Thus

$$N_s = (\lambda_1 + \lambda_2 + \lambda_3 + \lambda_4 + \lambda_5) T$$

is the total number of faults expected during the interval T, assuming that $\lambda_1, \lambda_2, \ldots \lambda_5$ are constant.

We now need to calculate the system reliability in terms of N_s, the number of faults expected during the operating period T. Since we are generally concerned with fairly reliable systems, N_s is small, and we may use the Poisson distribution described in section 2.14.

Equation (2.14) gives the probability of 0, 1, 2, etc., events occurring in terms of the expected number of events. The 'event' in our case is a system failure, and the reliability is the probability of zero events, that is, zero failure. This corresponds to the first term in the series and thus the reliability R is given by

$$R = \exp(-N_s) \ldots \tag{3.3}$$

Although this expression is not difficult to evaluate, it can in some circumstances be simplified. Using the exponential series, we have

$$\exp(x) = 1 + x + \frac{x^2}{2!} + \frac{x^3}{3!} + \ldots$$

Where x is small, we may neglect $x^2/2!$ and terms of higher degree, compared with x.

Thus

$$\exp(x) \approx 1 + x$$

and

$$\exp(-x) \approx 1 - x$$

Thus equation (3.3) reduces, for N_s small, to

$$R = 1 - N_s \tag{3.4}$$

This simplification is frequently used, and the condition usually imposed is that N_s should be less than 0·1. For if $N_s \equiv 0·1$, the reliability calculated from equation (3.4) is 0·9 and the second degree term $N_s^2/2!$ is equal to 0·005. Neglecting other terms gives the fractional error as

$$e_1 = 0·005 = 0·00555 \text{ or } 0·5555 \text{ per cent}$$

More exactly, the correct value of reliability is

$$R = \exp(-0.1)$$
$$= 0.90484$$

Thus the error involved in taking

$$R = 1 - N_s = 0.9$$

is

$$e_2 = 0.90484 - 0.9 = 0.00484$$

whence the true fractional error is

$$e_3 = \frac{0.00484}{0.90484} = 0.00535 \text{ or } 0.535 \text{ per cent}$$

Since the value $N_s = 0.1$ corresponds to a system having a reliability of about 90 per cent this error of just over 0.5 per cent is not serious. In the majority of situations it is less than the error expected in the reliability prediction due to uncertainties in the assumed failure rates λ_1, λ_2, etc.

A more discriminating measure of performance is obtained for small values of N_s by calculating P_s, the probability of system failure rather than the reliability. For the limiting case of $N_s = 0.1$ the accurate value of $P_s = 1 - R$ is 0.09516 compared with the approximate value of 0.1. This is an error of 5.08 per cent, which is greater than the error in reliability, but still likely be much less than the error caused by uncertainties in the data used for the calculation.

The approximation $R = 1 - N_s$ used above also implies that the unreliability, or probability of failure p, is $p = 1 - R = N_s$. Thus when N_s is small we may also equate the probability of failure to the expected number of faults.

3.4 Example of Series System

The following example shows how numerical data are used to estimate reliability. We consider an electronic unit containing the following components, with the failure rates shown

Component	Number used	Failure rate per cent per 1 000 h
Silicon diode	45	0·02
Silicon transistor	25	0·05
Resistor	90	0·005
Capacitor	25	0·01

The system failure rate is calculated by adding together the total failure rate for each class of component. In tabular form this is

Silicon diodes	45 × 0·02	= 0·9
Silicon transistors	25 × 0·05	= 1·25
Resistors	90 × 0·005	= 0·45
Capacitor	25 × 0·01	= 0·25 per cent
Total failure rate		= 2·85 per cent per 1 000 h

Thus

$$\text{the MTBF} = \frac{100}{2\cdot85} \times 1\,000 = 35\,100 \text{ h}$$

This estimate is likely to be somewhat optimistic, since we have made no allowance for the possible failure of a soldered joint or of any external connections via plugs and sockets. If we now suppose that the unit is used on a ship which makes voyages of about a month, what would be its reliability? In order to estimate this we assume that the equipment must operate for, say, 750 hours continuously until the ship returns to base where testing and repair facilities are available.

From the above calculation the expected number of faults in 1 000 hours is 0·0285. Thus the expected number in 750 hours is $n = 750/1\,000 = 0\cdot0214$.

Thus the reliability for each voyage, that is, the probability of no failure, is

$$R = 1 - n = 0\cdot9786$$

Here n is considerably less than 0·1, so that we can use the expression $R = 1 - n$ in place of

$$R = \exp(-\lambda)$$

with little error.

Since $1/0.0214 = 46.7$ we may also deduce that on average one voyage in about forty-seven would encounter a failure of the equipment. The calculation can be repeated for voyages of different durations with the following result

Duration in hours	250	500	750	1 000	1 500	2 000
Probability of system fault	0.00713	0.0143	0.0214	0.0285	0.0428	0.057
System reliability	0.99287	0.9857	0.9786	0.9715	0.9572	0.943

3.5 Parallel Systems

The systems we have considered previously have been basic or non-redundant, in other words they have contained no components other than those necessary to satisfy the functional requirements. Each component was essential for correct operation, and if any component failed the systems would not perform to specification. This is the type of system which meets the design specification at minimum cost and is thus the normal choice of the designer.

We now consider systems in which more than the minimum essential equipment is used, with the object of improving the system reliability. The general principle involved is to provide more than one way of meeting the functional requirements of the system, so that if one component fails, so disabling one subsystem, other subsystems can continue to operate satisfactorily. Such a system is called a 'redundant' system, and the designer's object is to achieve a substantial improvement in reliability at the expense of an acceptable increase in the amount of equipment used.

Redundancy can be introduced in a variety of ways, but a simple form as shown in figure 3.2 is the use of a number of alternative subsystems operating simultaneously, each capable of satisfying the functional requirements, with some switching mechanism at the output terminal which enables faulty subsystems to be disconnected. A simple example of this kind of arrangement would be an ultra-reliable d.c. supply for the safety system of a nuclear reactor. Here several d.c. supplies from a mains-energised rectifier, a float-charged battery, a stand-by d.c. generator, etc., may all be connected to a bus bar via power diodes. Should any source fail, it is automatically disconnected by the diode in series with it.

(a) Parallel reliability model

(b) Parallel connection of switches

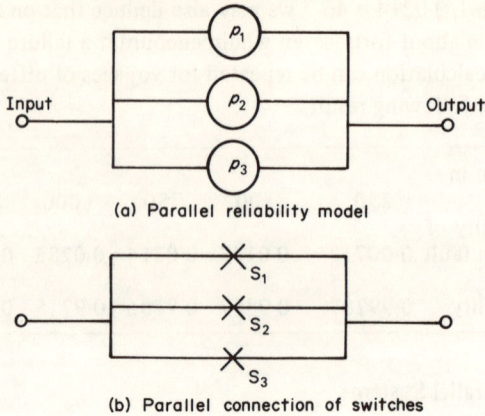

Fig. 3.2 Parallel systems

For a simple analysis we neglect the probability of the diode failing to disconnect a faulty unit. (In practice two or more diodes in series may be used to diminish this risk.) We then have three parallel subsystems, each of which can satisfy the operational requirements.

In this situation it is simpler to calculate the probability of failure than the reliability directly. We are given that any one of the three subsystems above is adequate, so that a system failure can only occur when all three of the subsystems have failed.

Thus the probability of this is

$$P_s = p_1 \times p_2 \times p_3$$

where p_1, p_2, p_3 are the probabilities of failure in each of the three subsystems in some specified interval of time.

The system reliability, assuming that p is small, is then

$$R = 1 - P_s = 1 - p_1 \times p_2 \times p_3 \tag{3.5}$$

For n identical systems, each having failure probability p_1, the system reliability is

$$R = 1 - (p_1)^n \tag{3.6}$$

The results given here are in practice an overestimate of the system reliability, since no account has been taken of the possibility of a fault in the switching mechanism. The effect of this is discussed in sections 3.13, 3.14 and 3.15.

A further consequence of parallel redundancy is that it may impose some electrical constraints upon the system design and thus make the attainment of the required performance more difficult.

3.6 Other Parallel Configurations

The previous section was concerned with a parallel system in which one subsystem alone could meet the functional needs. In a number of other situations, more than one subsystem may be needed. In general we may have n parallel subsystems, and require r of them to be effective for the system to function. This is sometimes called an 'r-out-of-n' system. As an example, a four-engined aircraft cannot normally fly on a single engine, but can do so with two engines or more. It is thus a 2-out-of-4 system.

If we consider such a system with four identical units in parallel, such that any two units can provide a working system, the condition for a system failure is that either three or four engines are inoperative.

If p is the probability of an engine failing, and this is independent of the failure of other engines, the probability of three engines failing and one engine being operative is

$$P_1 = 4 \times p^3(1-p)$$

Here p^3 is the probability that three engines will fail simultaneously, and $(1-p)$ the probability that the remaining engine will not fail. There are four ways in which this situation can occur, since any one of the four engines can be the operative one.

The probability that all four engines will fail is

$$P_2 = p^4$$

Thus the total probability of system failure is the sum of p_1 and p_2, that is,

$$P_s = 4p^3(1-p) + p^4 = 4p^3 - 3p^4$$

The reliability, that is, the probability of this not happening, is

$$R = 1 - P_s = 1 - 4p^3 + 3p^4$$

3.7 Example of Parallel System

As an example of parallel redundancy, consider a generating system which uses units having a mean time between failures of 5 000 hours.

What will be the reliability of the system for a 500 hour operating period if there are five identical units, any three of which can supply the required load?

The conditions imply that if three, four, or five machines fail, the system will fail. These situations are mutually exclusive, so that the combined probability of any one of the three situations occurring is given by the sum of the three separate probabilities.

Let p be the probability of failure of one machine during the 500 hour interval. We can select three machines from five in

$$\frac{5 \times 4 \times 3}{3 \times 2 \times 1} = 10 \text{ ways}$$

Thus the probability of having three faulty machines and two working is

$$P_3 = 10p^3(1-p)^2$$

Similarly, the probability of having four faulty machines and one working is

$$P_4 = 5p^4(1-p)$$

Finally, the probability of having all five faulty is

$$P_5 = p^5$$

Thus the total probability of failure is

$$P_s = 10p^3(1-p)^2 + 5p^4(1-p) + p^5$$
$$= 10p^3 - 20p^4 + 10p^5 + 5p^4 - 5p^5 + p^5$$
$$= 10p^3 - 15p^4 + 6p^5$$

The mean time between failures for a single machine is 5 000 hours. Thus the probable number n_f of failures in a 500 hour period is $500/5\,000 = 0.1$. If we use the approximation that this is equal to the probability of a failure, we can use a value $p = 0.1$ in the above expression for P_s, taking $n_f = p$.

This gives the probability of a system failure as

$$P_s = 10 \times (0.1)^3 - 15 \times (0.1)^4 + 6(0.1)^5$$
$$= 0.01 - 0.0015 + 0.00006$$
$$= 0.00856$$

whence the probability that the system does *not* fail, that is, the reliability, is given by

$$R = 1 - P_s$$
$$= 0.99144$$

The value $p = 0.1$ has been suggested as the greatest value for which the simplification $1 - R = P$ can be used. As an illustration of the error involved in the simplification, the same calculation follows with the exact value of p in terms of the expected number of faults, that is,

$$1 - p = \exp(-n_f)$$

Here

$$n_f = 0.1 \quad \text{and} \quad 1 - p = \exp(-0.1) = 0.90484$$

Thus

$$p = 0.09516$$

Substituting this in the expression $P_s = 10p^3 - 15p^4 + 6p^5$ gives

$$P_s = 0.00743 \quad \text{or} \quad R = 0.99257$$

The reliability predicted by the approximate method is 0.99144, thus the error involved is only 0.00133, or 0.134 per cent of the accurate value. This is a small amount, but it is significant since the reliability is high. A more instructive indication of the system performance may be obtained by considering the probability of failure, that is, $1 - R$. For the approximate calculation, the figure is $P_s = 0.00856$, compared with the exact figure of 0.00743. The error here is just over 15 per cent. The comparable error when $n_f = 0.1$ for a non-redundant series system is shown in section 3.3 to be just over 5 per cent.

There is a considerable discrepancy between the errors caused by using the approximation $p = n_f$ in the two cases. The reason for this lies in the expression for determining the system failure probability from the component failure probability p. In the non-redundant series system the component probabilities are simply added so that the system failure probability is a linear function of p.

For the parallel redundant case, however, the dominant term in the system failure probability involves p^3. This means that an error in p will have a greater effect on system failure probability than it does in the case of a linear relation.

Thus the widely used approximation which allows n_f to be equated to p if $n_f < 0.1$, although acceptable for a series system, should not be used for redundant systems without investigating the size of the error which it introduces.

3.8   Analysis of Mixed System

Many engineering systems can be represented by a combination of series and parallel sections. This arrangement is frequently used where some part of the system is particularly prone to failure and is consequently duplicated or triplicated. A typical reliability diagram for such a system is shown in figure 3.3. Here the third element, being much less reliable than the other elements, has been duplicated. The values $P_1 - P_5$ are the probabilities of failure of the various components of the system in a given interval.

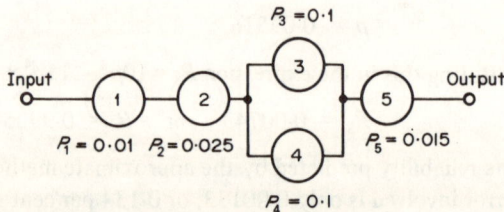

Fig. 3.3 Mixed series/parallel system

The duplicate pair of elements will cause a system failure only if they are both faulty. Assuming that they operate quite independently, so that the failure of one makes no difference to the probability that the other will fail, the probability of their joint failure is

$$P_6 = P_3^2 = 0.01$$

Thus the system can be replaced by a solely series system comprising P_1, P_2, P_6, P_5, and having system failure probability of

$$P = 0.01 + 0.025 + 0.015 + 0.01 = 0.06$$

Thus the probability that the system will *not* fail, that is, the reliability, is

$$R = 1 - P = 0.94$$

This compares with 0.85 for a series system in which unit 3 is not duplicated.

3.9 The Use of Redundancy

The analysis of parallel systems in section 3.5 and 3.6 shows that one method of improving the reliability of a system is to introduce additional equipment so that the overall system can continue to operate satisfactorily despite a failure of one or more elements of the system. This is one method of introducing redundancy into a system and thus increasing the reliability. In general terms a redundant system may be defined as one which has more than one way of producing the specified output.

The two or more methods of satisfying the requirements may involve sending the same information at different times along one path (temporal redundancy) or sending copies of the same signal at the same time along different paths (physical redundancy). In both cases there is a sacrifice of some system capacity on account of the repeated transmission of information. We are thus improving reliability at the cost of a decrease in system capacity, or the provision of additional equipment for the same performance, compared with a fault-free non-redundant system.

In many circumstances, however, the improvement justifies the cost.

An engineer faced with the need to improve the reliability of a given system has in general two options. He may examine the systems with a view to using higher quality components, improving the environment in which the system operates, and taking other steps which increase reliability. If these measures are either insufficient, or too costly, a redundant system may be the preferred solution.

The use of digital computers in system and process control provides an example of these two methods. For industrial applications to plant or processes which normally work for long intervals without a halt for maintenance, and which are expensive to run, a computer failure generally involves a serious loss of production. To reduce the likelihood of this, duplicate computers with provision for very rapid change-over are often used. Since the computers are standard items in large scale production, their maintenance and the provision of spares poses no difficulty. It is usually possible to test and if necessary repair one computer while the other operates the plant. In this situation the duplicate arrangement is cheaper than a specially designed ultra-reliable computer which would be non-standard, and permits some maintenance work to be done without interrupting the plant. The duplicate system of course requires additional space and power supply, but these represent a very small addition to the space and power already required for the plant.

The situation is quite different for the computer needed to control a manned space vehicle. Here again very high reliability, generally for a short period, is required, but space, power supply and weight are very restricted. Thus duplicate computers are not acceptable, and it is necessary to develop an ultra-reliable computer. This incorporates special high reliability components with redundancy applied to certain critical parts of the machine. This solution is much more expensive than the provision of two standard computers, but given the constraints on size, weight and power is the only possible one.

Redundancy is widely used in systems which are concerned with human safety particularly in passenger transport. In some cases the reliability may be specified, in others the redundancy may be required by law or regulation. For example, any automatic landing system used on British civil airlines must have a probability of serious failure not exceeding 10^{-7} per landing. This is roughly an order of magnitude better than the performance of an average pilot in favourable conditions, and so small that a redundant system is essential.

Before much data was available on failure rates and system reliability, it was much simpler for a regulating authority to specify the degree of redundancy needed. Thus all passenger vessels above a certain size were required to carry two independent radio installations, and all passenger aircraft on scheduled flights were required to have at least two engines.

3.10 Temporal Redundancy

The general use of temporal redundancy is to improve the reliability of systems which incorporate noisy or intermittent channels or circuits. The major application is thus to radio or long line circuits which may suffer from temporary fading, interference, or distortion. The method of introducing redundancy involves sending additional information either to enable the receiver to check that an error has occurred, or to correct it. This involves some form of information coding, in which additional signals are inserted in the transmission path. In telephony the inherent redundancy of normal speech allows many errors and uncertainties to be corrected, and a similar process can be applied to plain language text sent by telegraphy. Where certain names or figures are important they are usually repeated as a check on accuracy.

However, where the signals being transmitted contain no inherent redundancy, such as telemetry data or similar numerical information, it is necessary to insert additional check characters or digits.

A simple scheme used for binary coded information in digital form
is called 'parity checking'. If odd parity is used, each block of data
transmitted, typically a 7 bit or 8 bit character or a 16 bit computer
word, has one extra bit attached, chosen so that the total number of
ones in the block is an odd number. Thus if the data character is
01101010, the additional parity bit is 1, making five ones in all.

This procedure will reveal the presence of any single error, or odd
number of errors in a block, but will not detect two, four or an even
number of errors. The normal procedure when an error is detected is
to ask for the block, or the message which includes the block to be
repeated. Thus we expect all single errors within a block to be detected,
but not double errors. In a practical situation the probability of more
than two errors per block is usually negligible. The task of the system
designer is thus to reach an acceptable compromise about the size of
the block. If this is too large very little data transmission capacity is
sacrificed, but the probability of undetected errors becomes unacceptably large. On the other hand, if the blocks are too small, the detection
is efficient, but a considerable proportion of the channel capacity is
required for transmitting parity digits in place of data.

The situation can be quantified fairly simply if it is assumed that
an error in one bit has no effect on the probability of an error in
neighbouring bits. If the probability of any one bit being wrong is
p, and the block, including the parity bit, has length n bits, the probability of a single error in the $(n-1)$ data bits is

$$P_1 = (n-1)p$$

There are $n!/[(n-2)!\,2!]$ ways of selecting two bits from n, so that
the probability of two simultaneous errors, which will not be detected,
is

$$P_2 = \frac{n!}{(n-2)!\,2!}\,p^2$$

As an example of the use of parity checking we will examine the
character coding on a teleprinter. A widely used code has seven binary
digits to represent each character. Taking the probability of a single
digit being in error as 2×10^{-5}, the probability of a character being in
error is

$$P_1 = 7 \times 2 \times 10^{-5} = 1\cdot4 \times 10^{-4}$$

The use of a single parity bit gives an 8 bit block, with the probability of two bits being in error given by

$$P_2 = \frac{8!}{6! \, 2!} \times (2 \times 10^{-5})^2$$
$$= 28 \times 4 \times 10^{-10}$$
$$= 1.12 \times 10^{-8}$$

This occurrence will not be detected by the parity checking circuit. Thus the character error rate has been reduced by a factor of nearly 10^{-4} by the inclusion of parity checking. This involves a small amount of equipment at each end of the transmission channel, and a sacrifice of about 6 per cent of the channel traffic capacity, to transmit the parity bit and the occasional repeat of a block with a detected parity error.

In this example the probability of four bits in a block being in error is about 10^{-18}. Consequently, although any even number of errors would be undetected, the only case that need be examined is that of two errors.

There are some circumstances in which even a simple parity check can detect most double errors. These occur in digital magnetic tape units such as are used for backing stores on a digital computer. Multi-track recording is used, typically seven or nine tracks, including parity, a parity digit being added transversely across the tape for each character, and longitudinally along each track in a block. Thus a double error in a character would give a parity error along each track containing the error bits, but not within the character.

Parity generation and checking can be carried out by a sequential process, examining each bit in turn, or by performing a logical operation on all bits in the character simultaneously. In the first method, a binary counter is set initially to 1, for odd parity, or 0 for even parity. Each bit of the character is then gated to the counter input; a 0 will cause no action, a 1 will change the counter output from 1 to 0 or 0 to 1. The output of the counter at the end of the operation is equal to the parity digit required. For use as a checker, the parity digit so computed is compared with the parity digit received. Any disparity signifies an error.

A pyramid of two-input logic circuits may be used for the simultaneous calculation of parity. For even parity, those can be exclusive — OR gates, or modulo − 2 adders. If the inputs are A and B, the output is

$A . \bar{B} + \bar{A} . B$. Each adjacent pair of digits is connected to the inputs of such a circuit. The outputs or adjacent pairs of circuits are connected to the inputs of another similar circuit, and the process is repeated until a single output is obtained. This arrangement is most economical when the number of digits is a power of 2.

Integrated circuits are now available which compute the parity digit for an 8 bit character, both odd and even parity outputs being provided. A number of these may be used in combination to handle the parity digit for a larger block such as a computer word of 16, 32 or more digits.

3.11 Error-correcting Codes

A major requirement for successfully using parity checking for data transmission is some means of signalling back to the sender that an error has been detected and that the block in question must be retransmitted. This involves little difficulty for short distance links, particularly by line, but becomes expensive for point to point radio circuits and impossible for broadcast radio in which there is one transmitter and very many receivers.

A useful procedure in such situations is the use of error-correcting codes. There are very many of these, which generally require more redundant digits than parity checking.

The simplest versions will correct any single error in a block and detect most double errors. More complex codes will correct any two errors, and detect most triple errors, and require much more redundancy. Unfortunately the codes which are theoretically the most efficient become very expensive to implement for large block sizes, so frequently less than ideal efficiency is accepted in order to simplify the coding and decoding equipment.

One such set of codes are the Hamming codes, which provide enough information from check digits to correct any single error in a block. For small block sizes of 7, 15 and 31 binary digits, the codes may be described as 4 + 3, 11 + 4 and 26 + 5. The first number gives the number of information digits, and the second the number of check digits. Clearly the ratio of check to information digits decreases as the block size is increased, thus giving a more efficient arrangement in that the fraction of check digits is smaller. However, the larger the block the

larger the probability of an uncorrected double error, so that data on
the probable error must be obtained before one can choose a suitable
block size for a given situation.

As an example of the use of Hamming codes, we consider the 4 + 3
code. The check digits are always placed in location 1, 2, 4, etc., in the
block, so that the complete block comprises the seven bits:
$C_1 C_2 d_1 C_3 d_2 d_3 d_4$ where C_1, etc., are check digits and d_1, etc., data
digits.

The parity equations for determining C_1, C_2, C_3 are

$$C_1 \oplus d_1 \oplus d_2 \oplus d_4 = 0$$

$$C_2 \oplus d_1 \oplus d_3 \oplus d_4 = 0$$

$$C_3 \oplus d_2 \oplus d_3 \oplus d_4 = 0$$

Here the symbol $\oplus$ means addition to the base 2 (modulo 2), so
that all multiples of 2 are disregarded, and only the remainder is taken.
Thus

$$3 \oplus 5 = 0 \quad \text{and} \quad 2 \oplus 3 = 1$$

If we take the data digits $d_1 d_2 d_3 d_4$ as 0110 the parity equations give
$C_1 = 1$, $C_2 = 1$, $C_3 = 0$. Thus the complete block is 1100110.

If we now imagine that an error has occurred in transmitting the third
digit, say, the received block will be 1110110.

If the three above expressions are now evaluated, $C_1 \oplus d_1 \oplus d_2 \oplus d_4$
becomes $1 \oplus 1 \oplus 0 \oplus 0 = 0$. Similarly the other two expressions have
value 1. Thus the three expressions yield the binary number 011 or 3.
This indicates that the third digit is in error, and that the correct version
of the block is 1100110. The same indication of the bit which is in
error is given by the other versions of the Hamming code.

A number of other codes and parity checking methods are used, but
in all cases the more powerful the error-detecting and error-correcting
capacity, the greater the fraction of the block which must be allocated
to check digits rather than information. Thus there is always some loss
of information handling capacity when these codes are introduced.

The application of error-checking codes and similar methods of
improving the reliability of a single information channel is restricted to
situations in which the great majority of errors are transient ones which
affect only one bit in a character. They are incapable of improving the
system performance when a fault persists. The majority of electronic
component faults can be classed as permanent rather than transient,

since the characteristics of the component change irreversibly (for example, a coil may fail open-circuit, or a transistor fail short-circuit). The only method of maintaining a working system despite permanent faults is to provide some additional equipment to take over from a faulty component or sub-unit. This is the technique of physical or multichannel redundancy, in which the required output from a system can be obtained by following any one of several paths through the system. The system is so designed that if more than a minimum number of paths are working, the system output will be correct.

3.12 Multichannel Redundancy

Multichannel redundancy can be broadly divided into two types, active and passive. In active redundancy some testing or monitoring unit is used which checks the performance of the separate channels, and switches a good channel in place of a faulty one when it detects a failure. An example of this is the technique of diversity reception in long-distance radio transmission. It has been established experimentally that the pattern of fading at two receiving aerials spaced more than ten wavelengths apart will be substantially independent. Generally three spaced aerials are used, and after amplification a combining unit automatically selects the largest of the three signals for subsequent processing.

Since the probability that at any given moment all three aerials will receive a very small signal is much smaller than the probability of a single aerial receiving an equally small signal, the use of diversity reception gives much more reliable communication than is obtained with a single receiver.

Diversity reception will not give any improvement unless the aerials are spaced so far apart that the three signals appear independent of one another. In other words, the fact that the signal from one aerial is at the trough of a fade, and so very small, makes no difference to the signals expected at the other two aerials.

For example, if triple diversity reception is used, and a single channel receives a signal too weak to be usable for 10 per cent of the time, the probability that all three channels will receive an unusable signal at any given moment is $P = (0.1)^3 = 10^{-3}$.

Thus the probability of a complete system failure is 10^{-3}, and the system reliability is increased from 0·9 for the single channel to 0·999 for the triple system. This calculation assumes that the three channels are independent, and that the only likely failure is due to propagation conditions. Should there be a significant probability that communications could be interrupted by a failure of the single transmitter, or by interference from a nearby source of noise, the above calculation would be invalid. Diversity reception would offer no advantage against these faults, since they would affect all channels equally. Thus a more realistic assessment of the probability of system failure would be $P = (P_1)^n + P_2$. Here P_1 is the probability of failure of any one of the multiple channels, n is the number of channels, and P_2 is the probability of a fault which affects all channels simultaneously. The calculation of P_2 must include also the equipment used to select the best signal and to provide any subsequent processing of the signal.

Diversity radio reception has been used for many years and is an example of parallel redundancy which requires very little equipment to implement. In the simplest scheme the sole criterion used to determine which signal to connect to the outgoing circuit is the magnitude of the received carrier, which can easily be extracted from the detector circuit of the receiver. In more elaborate schemes the ratio of signal to noise in each channel is measured, and the channel with the best ratio is used.

In other parallel redundancy schemes there is no simple method of determining the state of the circuits by examining the received signal, and in such cases a pilot signal is used.

This is generally a continuous low level signal added to the channel at the transmitter and extracted at the receiver. To avoid causing interference to the information being handled it is usually located just outside the frequency band used for transmitting information. The amplitude of the pilot signal is continuously checked at the receiving station, and, where they are used, at intermediate amplifier stations along the route.

Thus, for example, in multichannel telephone circuits the pilot tone is used to adjust amplifier gain and equalisation to compensate for changes in line attenuation, but when the received pilot signal falls below a given level the system will automatically switch over to an alternative circuit.

The pilot signal maintaining process is applicable to analogue systems only; for digital systems some alternative method is required which will detect a faulty channel and switch one to a good one. One method is to send a prearranged signal at intervals, and incorporate a suitable detecting circuit at the receiver. This is difficult to arrange unless the possible set of information signals is restricted so as to exclude the signal used for testing. A simpler method is to send the same signal along all channels, and have some equipment at the receiver to detect a lack of agreement between channel outputs.

Various methods can be used to decide the course of action when there is disagreement, but a simple and effective procedure is to accept a majority decision. In a triplicate system this involves neglecting any channel which disagrees with the other two channels.

Multichannel systems can provide either active or stand-by redundancy. The former describes the situation when all channels are operational, and some switching mechanism selects a good one. In stand-by redundancy only the working channel is energised, and not until this fails is another channel switched on and connected into circuit. The advantage of stand-by redundancy is that equipment which is not switched on has a much lower failure rate than equipment which is working. Thus there will be fewer faults in a stand-by redundant system, and a longer time between failures, compared with an active system.

However, in the interval during which the fault is recognised and the oncoming channel is switched on and made ready for operation, no service will be available from the system. In some applications this break in service cannot be accepted, and active redundancy is required. An example of a system which takes some time to prepare for operation is a stand-by digital computer. After power has been applied, the working program and data must be loaded. If the consequent delay is unacceptable, the stand-by machine must be permanently energised. This is usually given some relatively unimportant task to perform, which it can jettison very quickly. The program and data needed for the stand-by role can either be resident in the machine, or available to be loaded from a high-speed device in a few milliseconds.

Another example occurs in the use of alternative power supplies in the event of a mains failure. In many situations, for example, a small radio transmitter, the break of a minute or less in service which would occur while a petrol-engined generator set was being started would be acceptable. However, in a large computer installation, a break of this duration without warning could easily ruin a job lasting a few hours, and would not be acceptable. In such circumstances a 'no-break' set would be needed, or an active redundancy scheme with the capacity to continue service during the start-up interval of the stand-by generator, possibly using secondary cells.

3.13　The Level of Redundancy

In the discussion of multichannel redundancy we have considered the result of replicating complete channels of equipment. Thus if we are concerned with the reliability of a radio transmitter, the provision of a second channel involves a completely separate and independent transmitter. Many electronic systems are assembled by interconnecting separate units, and in such cases the provision of a complete additional channel may not be the most effective way of introducing redundancy.

This can be shown by considering a duplicate system used to improve the reliability of the four element basic system shown in figure 3.4.

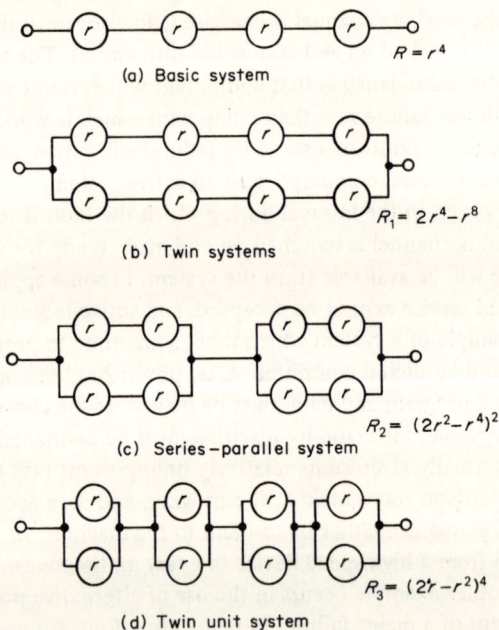

(a) Basic system $R = r^4$

(b) Twin systems $R_1 = 2r^4 - r^8$

(c) Series–parallel system $R_2 = (2r^2 - r^4)^2$

(d) Twin unit system $R_3 = (2r - r^2)^4$

Fig. 3.4　Schemes of redundancy at various levels

For simplicity we assume that the reliability of each element is equal to r. Then the basic system comprising four elements in series has a reliability $R = r^4$. The reliability of the duplicate system is thus

$$R_1 = 2R - R^2 = 2r^4 - r^8$$

Thus the reliability of a duplicated half of the system is

$$R_h = 2R_2{}^2 - R_2{}^2 = 2r^2 - r^4$$

The complete series-parallel system comprises two duplicated half-systems in series. Its reliability is thus the product of the reliabilities of the two separate half-systems, and is

$$R_3 = (2r^2 - r^4)^2 = 4r^4 - 4r^6 + r^8$$

Finally, for the arrangement in which each element is duplicated separately, the reliability of such a pair of parallel elements is $2r - r^2$.
Thus the system reliability, due to four pairs of elements in series, is

$$R_4 = (2r - r^2)^4$$
$$= r^4(2 - r)^4$$

What we have changed here is the 'level' of the duplication, that is, the size of the unit being duplicated. In the first case, the complete system is duplicated, and this corresponds to the highest possible level. The series-parallel case is intermediate, and the twin element case is the lowest level we can provide without enquiring into the component parts of each element. In general, the lower the level of redundancy the greater the effectiveness. The ultimate stage is reached when each component is duplicated, but this level is largely unattainable on account of the difficulties it brings to circuit design and the degradation in performance.

Some indication of the effectiveness of the three schemes is shown by assigning a range of values to r, as shown in table 3.1.

The figures in the table show clearly that the reliability is greatest when the duplication is introduced at the lowest possible level. In general

Table 3.1
The effect of duplication level on reliability

Value of r	0·7	0·8	0·9	0·95
Reliability of basic system	0·2401	0·4096	0·6561	0·8145
Reliability of twin channel system	0·4226	0·6514	0·8817	0·9656
Reliability of series-parallel system	0·5474	0·7576	0·9291	0·9811
Reliability of twin element system	0·6857	0·8493	0·9606	0·9900

this is impractical down at component level, but is possible at circuit level. Thus we can arrange duplicate or triplicate amplifiers and logic gates without incurring excessive penalties in circuit degradation. The problems of applying redundancy at this level are discussed in greater detail in chapter 5.

A more important point in relation to figure 3.4 is that we have taken no account of the error detection and switching mechanisms which are required for automatic change-over from a faulty unit to a working one. In some circumstances this may be justified; for example, if we consider the electrical power system used aboard a passenger liner, the engine-room is normally manned and in the event of one generator failing there is always someone on duty to start up the stand-by set.

In general, however, any system which requires human intervention to return it to service is considered to have failed, and the fault detection and switching process must be carried out automatically. Thus a more realistic reliability diagram for the duplicate systems is shown in figure 3.5, in which the error detecting and switching units are shown as having a reliability r_s.

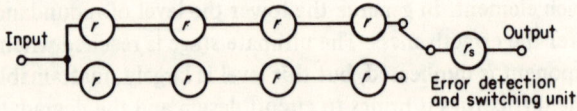

(a) Twin systems

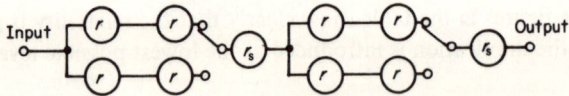

(b) Series—parallel system

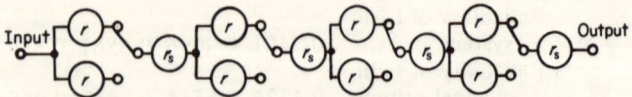

(c) Twin unit system

Fig. 3.5 Duplicate schemes with imperfect switching units

The previous expressions for system reliability need modifying to allow for the switching units, giving the following results

$$\text{Basic system as before} \quad r^4$$

Duplicate systems $\quad R_1 = (2r^4 - r^8)\, r_s$

Series-parallel system $\quad R_2 = (2r^2 - r^4)^2 \times r_s{}^2$

$$= (2 - r^2)^2 \times r^4 \times r_s{}^2$$

Duplicate element system $R_3 = r^4(2 - r)^4 \times r_s{}^4$

In this situation we can no longer predict that the lower the redundancy level the better. For as the level is reduced, the number of switching units in series increases, and so the reliability may reach a maximum and then start to fall.

For example, if we assume that the reliability of the switching unit is $r_s = 0.99$, and repeat the calculation for $r = 0.9$ and $r = 0.95$ we obtain the following values of system reliability

Table 3.2
Reliability of different schemes of duplication with imperfect switching units

Element reliability r	0·9	0·95
Basic system as before	0·6561	0·8145
Duplicate channel system	0·8729	0·9559
Series-parallel system	0·9106	0·9616
Duplicate element system	0·9228	0·9510

3.14 Use of Duplicate Switching Units

It is clear from table 3.2 that the benefits of redundancy when $r = 0.95$ may be partly nullified by the lack of reliability in the switching unit. Just as we have introduced redundant functional units to improve reliability, we can also introduce redundant switching units. The simplest method here is to duplicate the switching units, putting one into each channel.

We then revert to the three reliability diagrams of figure 3.4, but the unit r in the diagram now denotes a functional element together with its associated switching unit. Thus we can use the same expressions for reliability as before, but the element reliability r is replaced by the combined reliability of element and switching unit $r \times r_s$.

Thus the system reliabilities are as follows:

Basic system r^4

Duplicate systems $2(r \times r_s)^4 - (r \times r_s)^8$

Series-parallel system $(2r^2 \times r_s^2 - r^4 \times r_s^4)^2$

Duplicate element system $r^4 \times r_s^4 (2 - r \times r_s)^4$

For the two cases considered previously, with $r_s = 0.99$, the system reliabilities are as shown in table 3.3.

Table 3.3
System reliability with duplicated switching unit

Element reliability	0.9	0.95
Basic system	0.6561	0.8145
Duplicate systems	0.8633	0.9527
Series-parallel system	0.9168	0.9735
Duplicate element system	0.9533	0.9859

The table shows that with the individual unit reliabilities assumed in the example, there is no benefit gained from duplicating a single switching unit. However, when two or more are used (series-parallel or duplicate element systems) the overall reliability is improved. This is particularly evident for the duplicate element system with $r = 0.95$. Here the four switching units themselves have a reliability of only $(0.99)^4 = 0.9606$. This is little better than the reliability of a single functional element ($r = 0.95$).

A further deduction from table 3.3 is that the smaller the element which can be duplicated the better, provided that the switching unit is also duplicated. It can be shown that, given an overall basic system of a prescribed complexity and reliability, it is worth subdividing this system into separate units and duplicating each unit, until the reliability of the unit is of the same order as the reliability of the switching element. This assumes that there are no restrictions on size, cost or power supplies, and that maximum reliability is the only criterion of performance. Generally, many other factors are involved, and this ideal degree of subdivision is rarely attained.

With analogue circuits particularly, the signal monitoring and switching unit may be somewhat complex, so that the ideal functional element to be duplicated will be correspondingly large. With digital circuits on the other hand, simpler error-checking circuits are generally available, and thus smaller functional units are required for maximum reliability.

3.15 Triple Redundancy

Although duplication can afford a considerable enhancement to reliability, it requires some degree of complication in the monitoring and switching system if a faulty channel is to be detected rapidly and the spare channel switched into operation. Equally, if both channels are continuously operating, it is often difficult to be certain which is faulty if they disagree.

If, however, three channels are used, it is possible to compare them in pairs, and if two channels agree they are regarded as correct, and the third channel is regarded as faulty.

This process of taking a majority vote is simple to implement with digital circuits, and can usually be performed with somewhat greater complication for analogue circuits also. In either case there is no interruption in service during the switching operation, and for this reason triple redundancy is often used where such breaks cannot be tolerated.

For digital circuits the majority voting circuit needs to implement the logic function

$$X = A.B + B.C + C.A$$

where X is the output and A, B, C, the three inputs. This requires only three AND gates and one OR gate, or four NAND gates, and the voting circuit can thus be implemented in a single DIL package. With such a simple arrangement the reliability of a triple system will clearly be greatest when the redundancy is applied at a low level, thus requiring the basic system to be subdivided into small modules.

The result of triplicating a system with majority voting can best be analysed in terms of the probability of failure. We assume initially that the voting circuit is perfect. Thus the triple system will fail when either any two of the three channels are faulty, or all three. If the probability of a failure for a single channel is p, there are three ways in which two channels can fail, with the third healthy (probability $1 - p$). There is only one way in which all three channels can fail, so that the probability of a system failure is

$$P_s = 3p^2(1 - p) + p^3$$
$$= 3p^2 - 2p^3$$

If the voting circuit is imperfect, with a probability of failure of p_v, the total probability of system failure becomes

$$P_s = 3p^2 - 2p^3 + p_v$$

Unless p_v is very small, and less than $3p^2$, it becomes the dominant term in p_s and so mainly determines the system's performance. In this situation it is beneficial to triplicate also the voting circuit. The voting circuit and the channel may then be regarded as a single element in the reliability diagram, having a combined failure probability of $p + p_v$.

Thus the probability of system failure becomes

$$P_s = 3(p + p_v)^2 - 2(p + p_v)^3$$

For a system having high reliability and thus a small chance of failure the analysis in terms of system unreliability gives greater accuracy in computation, since these small numbers are handled directly. When dealing with reliability, we are concerned with the complementary probability which is very nearly equal to 1, and we need to use several more significant figures in the calculation to preserve the same overall accuracy.

However, the triplicate scheme with majority voting can also be analysed in terms of element reliability R, as shown in section 2.6. The reliability of a three element triplication scheme is

$$R_1 = 3R^2 - 2R^3$$

If the voting circuit has a reliability of R_v, the overall reliability is reduced to

$$R_2 = R_v(3R^2 - 2R^3)$$

since the triple element unit is in series with the voting circuit. If the voting unit is also triplicated, each channel becomes one element in series with one voting circuit, with a total reliability of

$$R_3 = R_v \times R$$

Thus when both voting circuit and element are triplicated, the overall reliability is the same as that of a triplicate system with each element having a reliability of $R_v \times R$ and is thus

$$R_4 = 3(R_v \times R)^2 - 2(R_v \times R)^3$$

3.16 Comparison of Duplicate and Triplicate Systems

We are generally concerned with systems which have high reliability and a low failure probability p. Thus in the expression $3p^2 - 2p^3$ for the failure probability of a triplicate majority voting system the dominant term is $3p^2$, and the term $2p^3$ is small in comparison. This compares with a failure probability of p^2 for a duplicate system (disregarding the fault liability of voting or switching circuits).

Since the triplicate system is more liable to failure and costs about 50 per cent more than the duplicate scheme it would appear clearly inferior. However, when the problems of arranging automatic change-over from a faulty channel to a working channel are considered in detail, the advantages of the triple majority voting system are realised. Generally, in a duplicate system the monitoring unit is complex and may require the insertion of a pilot signal. By comparison the majority voting circuit used in the triplicate scheme is usually much simpler and thus introduces much less fault liability, besides giving no interruption in service. Consequently in some circumstances the triplicate scheme may be preferable. A numerical example given later in the section illustrates this point.

On the other hand, if some interruption to service is permissible when a fault occurs and change-over switching is simple to arrange the duplicate scheme offers a cheaper method of improving reliability.

Some FM broadcasting stations are examples of the latter situation. It is not difficult to duplicate low-level parts of the equipment such as the master oscillator and modulator, and the remainder of the equipment can be arranged in duplicate, with both sets working. In the event of a failure in, for example, one power amplifier, the other power amplifier would provide a reduced service via its aerial feeder and half of the aerial array. Either manually or automatically the other half of the aerial array could then be switched to the remaining amplifier.

This would provide a radiated power only 3 dB below normal, if the switching were arranged to provide correct power matching. Since FM receivers contain amplitude limiters, the great majority of listeners would not perceive any change in programme level, but a few on the fringe of the service area would recognise an increase in the background noise level.

This arrangement gives a slightly degraded service during fault conditions, but as these occur only rarely a more lavish provision of spare equipment cannot be justified economically.

To show the effect of triplication we take the same basic system as used in sections 3.13 and 3.14, namely four elements in series each having a reliability of 0·9, and switching units with a reliability of 0·99.

Considering first the situation with perfect switching units, we have

a channel reliability of

$$R = (0.9)^4$$
$$= 0.6561$$

for four elements in series.

From equation (2.7) of section 2.6 the reliability of a duplicate majority voting system is

$$R_1 = 3R^2 - 2R^3$$
$$= 0.7268$$

For the series-parallel scheme of figure 3.6(b) each half-channel comprising two elements in series will have a reliability of

$$R = (0.9)^2$$
$$= 0.81$$

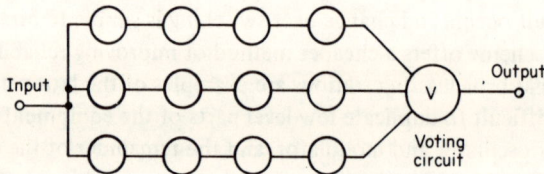

(a) Triplicate channel system

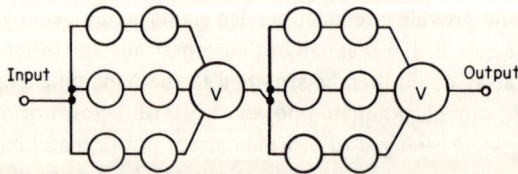

(b) Series-parallel system

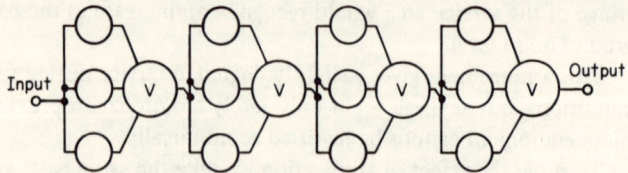

(c) Triplicated unit system

Fig. 3.6 Schemes of triplicate redundancy with majority voting

When triplicated this will have a reliability of

$$R_2 = 3R^2 - 2R^3$$
$$= 0.9054$$

The two sets of half-channels in series will have a reliability of

$$R_3 = R_2{}^2$$
$$= 0.8198$$

Finally, when each element is separately triplicated as shown in figure 3.6(c) the reliability for each triple is

$$R_4 = 3R^2 - 2R^3$$
$$= 0.972$$

since $R = 0.9$.

There are now four of these triples in series giving a system reliability of

$$R_5 = (R_4)^4$$
$$= 0.8926$$

As would be expected, the triplicate scheme has in this form a lower reliability than that of the duplicate scheme. The advantages of triplication become apparent only when the effect of imperfect voting and switching circuits is considered. If we take a reliability of 0.99 for the duplicate switching circuit, and 0.999 for the much simpler majority voting circuit, the triplicate scheme becomes preferable for a system with element reliability of 0.95, as shown in table 3.4, when each element is replicated as in figure 3.6(c).

It is clear from this table that the duplicate scheme is always more reliable than the triplicate scheme with majority voting if the possibility of a failure in the voting or switching circuits is disregarded. In practice this is quite unrealistic for electronic equipment, and if appropriate values of reliability are assigned to these circuits the advantages of triplication can be demonstrated. The figures of table 3.4 show that for the lowest possible level of redundancy, when each element is triplicated, this arrangement is more reliable than any form of duplication if the element reliability is 0.95 or more.

The reliabilities assumed here mean that the failure rate for the monitoring and switching circuit of the duplicate system is about ten times greater than that of the majority voting circuit. By implication,

Table 3.4
Comparison of duplicate and majority voting triplicate
redundancy schemes

Parameters of system	$r = 0.9$ Perfect switching circuit duplicate	$r = 0.9$ Perfect voting circuit triplicate	$r = 0.95$ Perfect switching circuit duplicate	$r = 0.95$ Perfect voting circuit triplicate	$r = 0.95$ Imperfect switching circuit duplicate	$r = 0.95$ Imperfect voting circuit triplicate
Multiple channels	0·8817	0·7268	0·9656	0·9095	0·9559	0·9086
Series-parallel	0·9291	0·8198	0·9811	0·9474	0·9616	0·9455
Multiple element	0·9606	0·8926	0·9900	0·9713	0·9510	0·9674

the switching circuit will be ten times more complex than the voting
circuit. In many situations the ratio may be much greater than ten, so
making the triplicate scheme more attractive.

Some more detailed consideration of the practical problems of
redundancy and the design of voting and monitoring circuits is given
in chapter 5.

3.17   Quadded Logic

Both duplicate and triplicate redundancy schemes incur penalties on
account of the unreliability of the switching or voting circuits which
are required for selecting the correct signal when the channels disagree.
This penalty could be avoided if there were no need for these additional
circuits.

One form of redundancy applicable to digital equipment which gives
the same order of reliability improvement as duplication or triplication
and needs no extra circuits uses four information channels. It is called
'quadded logic' and provides the error-correcting mechanism by
increasing the number of inputs to each gate and cross-connecting the
four channels as shown in figure 3.7.

The method is most effective when successive logic functions in
each channel are alternately AND and OR.

It can be shown that a single error will be corrected at the logic
element two rows further along the channel. However, if two errors

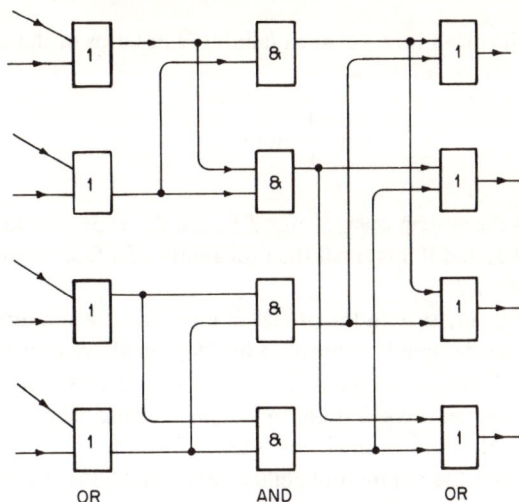

Fig. 3.7 Section of quadded logic network

occur close together, they may not be corrected. The major disadvantage of this scheme is the difficulty of testing. As with all schemes of redundancy, full benefit is obtained only if all elements are working properly at the start of the operating period. With triplication and majority voting, it is easy to introduce a fault detector at each voting unit which will warn the maintenance staff that there is disagreement between the channels and thus a fault which needs correcting.

With quadded logic some partial testing can be performed by splitting the power supply so that each channel can be energised separately and tested. This procedure is complicated and cannot test all possible states of the quadded logic array. For this reason quadded logic has not been used to any significant extent in the design of high reliability digital systems.

3.18 MTBF of Redundant Systems

In previous calculations we have assumed a failure rate which does not vary with time. This is a valid assumption for electronic components which generally have no 'wear-out' mechanism, and is discussed in more detail in chapter 4. The assumption leads to a very simple expression for the mean time between failures M in terms of the failure rate λ faults per hour, as

$$M = \frac{1}{\lambda} \text{ hours}$$

If the system operates for T hours, the expected number of faults is $n = \lambda T$, and if n is small the probability of a fault in the period T is $p = n = \lambda T$.

If in any system the probability of failure is not directly proportional to time the simple expression for M given above is invalid. This may arise if λ is a function of time, or in redundant systems. We are concerned with the second situation. In a duplicate channel arrangement the system fails only when both channels are in operative.

If p is the failure probability per channel, the system failure probability is p^2, which as a function of the operating period, is

$$p^2 = \lambda^2 T^2$$

Thus we can no longer use the expression $M = 1/\lambda$, and must use the more general expression giving M as a function of reliability R

$$M = \int_0^\infty R \, dt$$

(proved in section 1.10).

For a single channel with a constant failure rate λ we have $R = \exp(-\lambda t)$.

Thus

$$M = \int_0^\infty \exp(-\lambda t) \, dt = \left[-\frac{1}{\lambda} \exp(-\lambda t) \right]_0^\infty$$

$$= 0 - \left(-\frac{1}{\lambda} \times 1 \right) = \frac{1}{\lambda}$$

as before.

For a duplicate system the reliability of the system is

$$R_s = 2R - R^2$$

(R = channel reliability)

Thus the MTBF of the system is given by

$$M_s = \int_0^\infty (2R - R^2)\, dt$$

$$= \int_0^\infty [2\exp(-\lambda t) - \exp(-2\lambda t)]\, dt$$

$$= \frac{2}{\lambda} - \frac{1}{2\lambda} = \frac{3}{2\lambda}$$

This is only 50 per cent more than the MTBF of a single channel. However, if we are concerned with any reasonable level of reliability, the operating period will be much less than the MTBF, and the duplicate system will be markedly superior to the single channel. This is clear as the failure probability for the system is p^2, compared with p for a single channel. Thus the smaller the value of p, and so the more reliable the basic channel, the greater the improvement obtained by duplication.

For a triplicate scheme with majority voting

$$R_s = 3R^2 - 2R^3$$

$$M_s = \int_0^\infty [3 \times \exp(-2\lambda t) - 2 \times \exp(-3\lambda t)]\, dt$$

$$= 3 \times \frac{1}{2\lambda} - 2 \times \frac{1}{3\lambda}$$

$$= \frac{5}{6\lambda}$$

which is less than that of the basic channel.

If, however, the triplicate system is operated on a parallel basis, so that the system fails only when all three channels fail, the system reliability is

$$R_s = 1 - p^3$$

$$= 1 - (1 - R)^3$$

$$= 3R - 3R^2 + R^3$$

where the MTBF of the system is

$$M_s = \int\limits_0^\infty [3 \times \exp(-\lambda t) - 3 \times \exp(-2\lambda t) + \exp(3\lambda t)]\, dt$$

$$= \frac{3}{\lambda} - \frac{3}{2\lambda} + \frac{1}{3\lambda}$$

$$= \frac{1}{\lambda} + \frac{1}{2\lambda} + \frac{1}{3\lambda} = \frac{11}{6\lambda}$$

In general, for n parallel channels, only one of which need work for the system to be operative, the reliability is

$$R_s = 1 - (1 - R)^n$$

and

$$M_s = \frac{1}{\lambda} + \frac{1}{2\lambda} + \frac{1}{3\lambda} + \ldots + \frac{1}{n\lambda}$$

3.19 Approximate Methods

The methods described above for estimating system reliability are based on an assumption that the equipment being analysed is completely specified. In the preliminary stages of design this is generally not a valid assumption. Several different methods of satisfying the design requirements may be investigated and when the design is still at the outline stage some comparative assessment of the reliability and other characteristics of the various methods is required to decide which method should be used.

For this kind of reliability assessment we require a simple procedure which will be reasonably accurate when comparing one design with another, the absolute accuracy being less important.

A convenient basis for this is a count of active devices. For transistor circuits based upon discrete components it is found that for each transistor used in digital equipment there are on average about four diodes, two capacitors and five resistors. It is thus convenient to add the failure rates of all these components to that of the transistor to obtain a suitable value of total component failure rate per circuit.

Taking the same failure rates as used in the example of section 3.3, the total failure rate associated with each transistor in a digital system is found to be as follows

1 transistor	0·02 per cent
4 diodes	0·08 per cent
5 resistors	0·025 per cent
Total	0·125 per cent per 1 000 hours

Thus for a system containing 600 transistors, we have a total failure rate of

$$\frac{600 \times 0.125}{100} = 0.75 \text{ components}/1\,000 \text{ hours}$$

The estimated system MTBF is thus

$$\frac{1\,000}{0.75} = 1\,333 \text{ hours}$$

The number and type of components associated with each transistor depends, of course, upon the nature of the equipment being analysed. The figures given above are typical of discrete component logic circuits, using diode gating. This type of circuit is still used where interference is present, and large signals must be used to avoid errors. Most logic systems are now assembled from integrated circuits, and all that is needed for a preliminary assessment is a package count.

Communication systems, although using some integrated circuits, still incorporate many discrete components, and since some kind of filtering or tuning is essential, the components associated with each transistor must include perhaps 0·5 inductor.

Although this procedure involves some fairly crude approximations, it does nevertheless afford some guidance about the relative reliabilities of various system designs at an early stage in the design process.

3.20   Integrated Circuits

During the 1970s two developments have resulted in major changes in the way in which much electronic equipment is constructed. In order to improve resolution and calibration stability, and simplify data storage and transmission there has been a steady move to digital methods of measurement, data conversion and transmission. At the same time advances in microelectronics have enabled large numbers of transistors and their interconnections to be fabricated as integrated circuits on small slices of silicon.

Consequently nearly all equipment involved in these applications is now largely assembled from integrated circuits, so needing far fewer components, but of greater complexity. In high volume applications large scale integrated circuits such as microprocessors and their associated packages can be used, so reducing apparatus previously assembled from hundreds of discrete components and small scale integrated circuits to two or three packages.

A similar but less pronounced change has also occurred in linear circuits, as operational amplifiers, modulators, demodulators, multi-stage gain-controlled h.f. amplifiers, etc., are now available in integrated form. When assessing system reliability it is thus necessary to consider integrated circuits as another class of component, and allocate to them appropriate failure rates.

This can be done from records of equipment in service where the integrated circuits have been produced with only minor changes for a number of years. This situation now obtains for small scale and medium scale integrated circuits using TTL, ECL and MOS technology, and for early types of microprocessors such as the 8080 and its associated packages. More recent designs must be assessed by extrapolating from known data on earlier types, making some allowance for extra complexity, and from accelerated life tests. In general the results rather surprisingly suggest that failure rates are nowhere near proportional to circuit complexity, and in nearly all cases integrated circuit versions of equipment are more reliable than their discrete component predecessors.

This is mainly due to the elimination of a large proportion of joints and connections between components, and the replacement of tens or hundreds of components by a single integrated circuit. Thus if each of, say, a dozen integrated circuits has a somewhat higher failure rate than each of the five hundred components they replace, the equipment as a whole will be much more reliable.

Some useful data on the performance of transistors and TTL integrated circuits in a benign environment has been obtained from field records of a large computer manufacturer. These covered a range of equipment which accumulated about 10^{10} component-hours per year. The average failure rates were as follows.

Transistors — lowest figure for logic circuits 0.02×10^{-6}/h

— average for all switching applications 0.065×10^{-6}/h

Digital integrated circuits — small scale

— less then 13 gates 0.015×10^{-6}/h

— 13 to 50 gates 0.03×10^{-6}/h

— 51 to 200 gates 0.06×10^{-6}/h

Linear integrated circuits 0.1×10^{-6}/h

These records show that the circuits containing 51 to 200 gates have a failure 4 times greater than that of small scale integrated circuits, while having on average nearly 10 times more gates. The comparison is not an entirely fair one since the small scale circuits have been in production longer and would therefore have benefitted from a greater improvement in reliability during the production phase.

An interesting feature of one group of integrated circuits was the reduction in observed failure rate over a five-year period, which fell from an initial value of 0.5×10^{-6}/h to 0.025×10^{-6}/h. This is usually ascribed to the better control of the fabrication process and the greater understanding developed as production builds up.

Other data recorded a little later by a microprocessor manufacturer gives 8 failures during an operating time of 1.3×10^{8} device-hours. This corresponds to an observed failure rate of about 0.062×10^{-6}/h, for an early 8-bit microprocessor containing about 6000 transistors.

For critical applications integrated circuits can be given a 'burn-in' period and extra screening; one manufacturer quotes an inherent reliability after the infant mortality phase of 0.01×10^{-6}/h at 55 °C for small and medium scale integrated circuits which have been given a high level of inspection during the process of manufacture.

One particular difficulty which arises in all large scale integrated circuits is that of testing. Microprocessors in particular possess such a very large number of possible states (10^{30} or more) that to test every operation with all possible combinations of data would take an unacceptably long time. Thus a limited amount of testing must be tolerated as a commercial necessity. A further complication is that instructions requiring the interaction of external units such a storage packages may have a number of critical time delays all of which should ideally be tested under worst case conditions so as to establish adequate operating margins during normal use.

Even devices which are operationally much simpler, for example, large capacity dynamic stores may prove difficult to test, since in addition to a number of critical timing requirements, they may reveal pattern-sensitive faults. These are errors which occur only with particular patterns of data. Searching experimentally for these in a 64K store is an extremely slow process; it may be possible to identify likely failure patterns if the precise layout of the stores is known, but information of this kind is not generally disclosed by manufacturers.

3.21   Design Faults

Experience with complex equipment built from high quality components shows that unexpected behaviour and in some cases system shut-down occurs when subsequent tests reveal no component failures. Such events can occur when an attempt has been made to use the equipment in a manner not intended by the designer, or not specified by the purchaser when procuring the equipment. A similar event can occur when failure can be ascribed to a combination of input data and environment which the designer failed to consider or thought would never arise.

These faults are often called 'design faults' although they could more properly be called communication failures since in most cases they arise due to a failure in communication between the various people involved with the equipment. They may involve quite separate teams concerned with specifying requirements, evaluating tenders and placing orders, and the ultimate users. Where the teams are widely separated geographically there are clearly many opportunities for misunderstanding, and the reduction or elimination of design faults requires very careful project monitoring and management, good documentation, and very precise specification of requirements and performance at all stages of the operation.

There is of course no way in which the probability of this kind of failure can be predicted since it is in broad terms caused by a human error and not by any component failure.

A particular category of design faults which has grown with the increasing complexity of equipment has been called 'sneak circuits'. These may be defined as designed-in signal or current paths which cause an unwanted function to occur or inhibit a wanted function.

They generally arise because the designer has not taken into account transient circumstances which can arise but are not part of the steady-state environment. Examples may be switching on power supplies one at a time rather than simultaneously, or the inadvertent disconnection and re-connection of power supplies while equipment is working. One example was discovered only after 50 successful firings of a particular type of rocket booster, when an unexpected engine cut-off occurred. The cause was the reversal of the normal order in which the umbilical cords connecting the booster to the launching pad become disconnected during the launch.

Another cause is the lack of understanding of the designer of one module in a system about the complete flow of signals and power in all other modules connected to his. This sometimes arises when a particular module is redesigned and the new arrangement, being regarded as a minor up-date, is not subjected to adequate testing.

A solution to the problem now generally used in avionics is the analysis of all such possible sneak circuits by computer. This avoids a great deal of very tedious work in which humans can easily make mistakes, but computers should not.

Any computer solution of this kind, however, is no more reliable than the program which it uses, and there is clearly a need for exhaustive testing before the computer analysis can be relied upon.

An unfortunate example of a programming mistake occurred in a design program which incorrectly combined two stresses acting on pipe-work supports for nuclear reactors. The resulting design predicted supports which were of inadequate cross-section and so failed early in their service life. When the fault was identified as a design error, all power plants using the same reactor design had to be shut down for repairs. As there were about 20 involved this was a very expensive mistake.

3.22 Software Faults

A failure mode analogous to a design fault can occur in computer and microprocessor systems when there is an error in the program. This causes an incorrect action which does not arise from any component failure. Although the designer always hopes to eliminate these errors during program testing it is possible that the fault will be revealed only by a specific combination of data. It may thus lie dormant for perhaps years until the critical combination occurs.

In most microprocessor systems the program is held in a read-only store and is thus incorporated into a hardware component; in these circumstances the user cannot distinguish between a hardware fault which corrupts a correct instruction and a software fault which caused the loading of an incorrect instruction in the first place.

In this type of equipment it is necessary to compare the program as read from the store with the program which was originally loaded into the store and read back to check the loading operation. If these differ a hardware error has occurred — if not the cause is a software fault.

Clearly it is not possible to assess a program's reliability in terms of the failure rates of its components, as can be done for hardware. Experience suggests that as program size increases, the probability of error increases at a much greater rate. Thus if programs can be subdivided into fairly small units, perhaps written as subroutines, each of which can be tested separately, testing time can be saved and the resulting combined program is likely to contain fewer errors. This procedure does, however, require the interfaces between program modules to be specified and controlled carefully.

An additional difficulty is that the principles of constructing robust and reliable programs are less firmly established than those for designing reliable hardware.

Various methods of program design have been advocated which are claimed to reduce the probability of errors and to generate robust and reliable software. These include

- 'Top-down' program design
- structured programming
- the adherence to strict programming standards
- the use of high-level languages which closely regulate the interfaces between modules and the structure of programs.

Although these techniques all have their place, no single one can be regarded as a panacea, and the generation of error-free programs remains a difficult task requiring much care and effort.

3.23   Effect of Failure Mode on Design

The previous sections have dealt with components and systems which failed in an unspecified manner. Many components have a pronounced failure pattern and use can be made of this information to make more effective use of redundancy. At component level, elements which usually fail to a short-circuit should be replicated in series, and those which usually fail to open-circuit should be replicated in parallel.

Thus open relay contacts are much more liable to open-circuit failure due to dust and contamination preventing good contact. Consequently, standard telephone relays have twin contact sets operating in parallel for each circuit.

Power diodes are generally more susceptible to short-circuit failure, so that when used for paralleling several d.c. supplies to a load requiring continuity of supply, they should be replicated in series.

This approach at component level requires a knowledge of the separate failure rates to open-circuit and short-circuit, rather than the gross failure rates usually quoted. Some information of this kind has been assembled for the design of safety and alarm circuits, where failure to a short-circuit and failure to an open-circuit can have very different implications for the user. This point is considered in more detail in section 5.17.

PROBLEMS

1. The probability of each engine of a three-engined aircraft completing a flight without failure is 0·99. The probability of completing the flight with one engine inoperative is 0·45. If more than one engine is inoperative the flight is abandoned. What is the probability of the aircraft reaching its destination?   [0·9835]

2. A communication system comprises units A, B, C, D, E, and is normally unattended apart from monthly maintenance visits. The failure rates for each unit are as follows

Unit	A	B	C	D	E
Faults per 1 000 hours	0·003	0·0015	0·08	0·065	0·0045

Units C and D are duplicated with automatic switching so that the system is operative if either of the twin units is working. What is the reliability of the system, and how many unscheduled visits to repair faults are likely during a ten year period?   [0·9811, two or three visits (expected number = 2·264)]

3. An unmanned communications satellite incorporates a microwave repeater having a mean time to failure of 30 000 hours. Find the reliability for a two year operating period of (a) a single channel,

(b) two parallel channels, (c) three parallel channels. Assume that the
link is operative if one channel is working and that the reliability of
the switching unit is 0.99.   [(a) 0.5577, (b) 0.7963, (c) 0.9043]

4. The following table gives the numbers and failure rates of the
components used in a communications unit.

Component type	Number in system	Failure rate per cent per 1 000 hours
Transistor	28	0.015
Diode	52	0.008
Integrated circuit	30	0.02
Resistor	106	0.001
Capacitor	27	0.001
Miscellaneous	35	0.0015
Connectors	8	0.007

This unit is used with a packaged power supply unit having a MTBF
of 10^5 hours. Calculate the overall system reliability over a one year
period using (a) a single power unit, (b) duplicate power units operated
in parallel redundancy, assuming perfect switching.
[(a) 0.7914, (b) 0.8578]

4 Component Failure Data

4.1 Variation of Failure Rate with Time

The reliability calculations given in the previous chapter are based upon the assumption that the failure rates of the components do not vary with time. Although this assumption leads to the simplest prediction techniques it is permissible only if it represents fairly accurately the measured behaviour of components in a working system. In general terms the assumption may not hold for mechanical components, but is a good approximation for electronic components, particularly since the introduction of transistors in place of valves.

When electronic equipment was based upon thermionic valves, the general pattern of failures could be separated into three distinct regions, as shown in figure 4.1. This illustrates the so-called 'bath-tub' curve, divided into an early period of falling failure rate, a period of constant failure rate associated with the normal life of the equipment, and a final stage of increasing failure rate corresponding to the 'wear-out' phase at the end of life.

The first phase lasted typically for periods up to several hundred hours, and was often hidden from the customer since it occurred during testing and commissioning at the manufacturer's factory. It was caused by minor errors in the assembly of the equipment, imperfect joints, or a few sub-standard components not detected by previous testing.

The final phase was caused by components which had a finite life due to mechanical wear or physical changes such as the slow evaporation of materials which occurs in the oxide-coated cathodes of thermionic valves. In many systems the components liable to suffer from this type

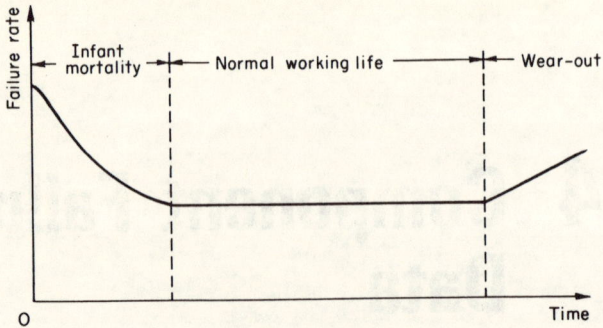

Fig. 4.1 Typical variation of failure rate with time

of failure could be identified and replaced before reaching the final phase of their lives. By this means the working life of the equipment could be prolonged much beyond the period attainable without replacement. Such a procedure is, however, possible only for equipment which can readily be maintained, and would, for example, be impossible for underwater repeaters on a transatlantic cable.

Experience with transistorised equipment generally shows a somewhat longer period of early 'infant mortality' when the failure rate is falling, and an approximately constant failure rate thereafter. Since transistorised equipment is usually many times more reliable than valve equipment, the number of failures per month is much reduced, and it is much more difficult to discern any long-term trend since the statistical fluctuations are rather large.

For equipment used in benign conditions, that is, well cooled, without vibration or shock, and with components stressed well below their maximum ratings, the failure rates are so low that component manufacturers cannot afford to test hundreds of thousands of components for long periods to obtain precise data. Most information thus comes from users of large equipment such as digital computers, electronic telephone exchanges or communications systems which are often operated continuously for several years.

If fault statistics from such systems are plotted as a function of time, they indicate generally that the failure rate is either constant with time, or slowly decreasing. Thus the assumption of constant failure rate can be regarded as valid, or if incorrect, slightly pessimistic. In either case it is acceptable for the purpose of reliability prediction.

The third 'wear-out' phase of figure 4.1 is largely irrelevant to transistorised equipment. Although it can be argued that every transistor can or integrated circuit package contains some minute amount of contaminant which will ultimately cause failure, all system experience suggests that the equipment concerned will be scrapped on account of technical obsolescence long before any wear-out phase is encountered.

The first phase of figure 4.1 is still encountered in many transistorised systems, but it is usually disregarded for reliability prediction as the major part of it may occur during the manufacturing or commissioning period. In order to ensure that this is so, many customers who are concerned with high reliability call for a minimum period of operation before equipment is delivered to them. This is intended to ensure that most of the early failures will have been repaired before the equipment goes into service.

4.2 Types of Failure

A component fails when its characteristics change to such a degree that it cannot operate to its specified level of performance. The type of failure may be classified in several ways, depending upon the magnitude of the change, and the rate at which it occurs.

A large and very rapid change is called a 'catastrophic' failure. It generally occurs as an open-circuit or short-circuit failure which is irreversible. Its other attributes are then 'permanent' and 'complete', and it is the consequence of a physical change in the component.

Examples of this kind of failure are open-circuits in wire-wound resistors or relay coils, or short-circuit failures in capacitors. The only cure for this type of failure is the replacement of the faulty component.

Catastrophic failures may, however, not be permanent; occasionally a short-circuit may be caused by a small whisker of wire which bridges two adjacent conductors, for example, on a printed circuit board. If the equipment is part of a fixed installation with no vibration or forced air circulation, the fault will persist until removed by maintenance action. But if the equipment were operated in a vehicle, subject to shock and vibration, this could dislodge the wire causing the fault. If the fault had caused no damage to other components, the equipment would resume normal operation, and the failure would be classed as temporary.

A similar situation may occur with short-circuit failures in some metallised film capacitors which suffer a short-circuit fault due to a breakdown of the dielectric. If these are used in a circuit which makes enough energy available to vaporise the conducting bridge through the dielectric, or the small adjacent area of conducting film, the small faulty area can be isolated from the rest of the capacitor, which can continue to operate. In favourable circumstances the voltage across the capacitor may not change significantly during the 'burning-out' of the fault, and no equipment malfunction may occur. Otherwise the burning-out process may cause a temporary change in capacitor voltage and so a temporary system failure. In the least favourable circumstances the capacitor may be in a circuit which cannot supply any appreciable energy, for example, a signal filter, in which case the short-circuit failure will persist and the fault will be classed as permanent.

This situation illustrates one of the complications of reliability analysis. The same physical event in the capacitor may or may not cause a system failure, depending upon the circuit function assigned to it.

The same situation is true for all components which suffer from 'degradation failure'. This occurs when the characteristics change slowly with time, at some point causing the circuit to operate incorrectly. The moment at which this occurs clearly depends upon the function of the component; a capacitor used in a very stable oscillator may be unacceptable if its value changes by 1 per cent, whereas a decoupling capacitor may decrease in value by 50 per cent or increase by 100 per cent without causing a circuit fault. In consequence it is necessary to define precisely the range of characteristics which are acceptable for a component so that there is no uncertainty about the point at which a degradation failure occurs.

For passive components such as resistors and capacitors this kind of definition is usually fairly simple. We may specify, for example, that the value of a resistor shall lie within the range 100 ohms ± 10 per cent, subject to a temperature which may vary between −40°C and +70°C. Devices such as transistors which have many more characteristics of interest to the circuit designer pose more difficult problems. Although it is fairly easy to specify the range permitted for each separate parameter of the device, in a particular circuit some extreme combinations of values may be unsatisfactory, and it requires much calculation or experiment on the circuit to determine the permissible limits when several parameters may vary simultaneously. A further classification of failure is between partial and complete failure. A catastrophic fault in a component generally causes a complete system failure, but a degradation fault may merely impair the system without completely preventing its operation. This is often the case with analogue systems, particularly if some noise or other disturbance is normally present, but is less likely with digital systems.

If, for example, we consider a radio link used for teleprinter operation, a substantial decrease in radiated power may occur in a transmitter left unattended for long periods. This would not cause a complete failure of communication, but the diminished signal-to-noise ratio would cause an increase in the error rate. Although the transmitter might be classed as faulty, since it was not radiating the specified power, the complete link would be classed as a partial failure only.

Developments in component design and technology in recent years have considerably improved their stability, so that degradation faults are becoming less frequent. They are, however, still significant in equipment which is used under very adverse conditions such as high temperature.

4.3 Factors Affecting Failure Rate

Component failure rates are affected by the mechanical, electrical and thermal environment in which they are required to operate. Shock and vibration have relatively little effect upon electronic components which are light, and can be encased in an insulating medium such as epoxy resin to give mechanical support and protection. Electromechanical items such as relays and contactors are much more susceptible to maloperation due to shock and vibration, and may need mounting on anti-vibration supports. The more vulnerable items are heavy components such as transformers which need very secure fixing, and in order to ensure that the windings are not displaced by shock, the complete transformer may be encapsulated in resin.

Environmental testing of electronic equipment shows that shock and vibration does increase the failure rate, but the majority of failures occur in electromechanical items, or from the mounting and interconnecting of electronic components rather than from inherent weakness in the components themselves.

As an indication of the levels of stress encountered, the acceleration measured in a sudden shock in a wheeled vehicle when a wheel bounces may be up to $40\,g$. Corresponding figures for aircraft take-off and landing are about $5\,g$, for ships in heavy weather up to $8\,g$, and missile launching up to $50\,g$. Continuous vibration is also a problem as it may excite a mechanical resonance and its effect will thus be multiplied.

The peak accelerations are usually lower than for the sudden shocks, for example, only about $1g$ at 5–15 Hz for a car body, about $2–3g$ at frequencies up to 25 Hz for ships, $5g$ at frequencies up to 500 Hz on aircraft, and up to $40g$ at 30–5 000 Hz in missiles.

Some experiments to test the mechanical strength of the welds inside metal-canned transistors indicated that centrifugal accelerations of around twenty thousand times g were required to produce a significant number of failures. There is obviously a considerable margin of safety when operating with accelerations of $40g$ and below.

4.4　The Effect of Temperature

One form of stress which reduces the reliability of all electronic components is high temperature. All test information shows that as the working temperature of a component rises above normal, the failure rate increases. In many cases the relation between failure rate and temperature agrees with theoretical predictions which relate the speed of a chemical reaction to temperature.

The argument for this is that many modes of failure which occur in electronic components arise from a chemical reaction between the material used for the component and some very small amount of contaminant. The contaminant may be trapped within the component during manufacture, or may leak or diffuse into it as a result of imperfect sealing between the component and the atmosphere. If sufficient contaminant is present, the chemical reaction between it and the component may ultimately cause the component to fail because the reaction consumes the material of the component, or because the reaction products damage the component. The first situation may arise when an imperfectly sealed wire-wound or metal film resistor is used in a humid environment. The moisture and the applied voltage provide the conditions for electrolysis, so removing part of the resistor material, and ultimately causing an open-circuit failure.

Examples of the second situation occurred in many early plastic mouldings which were subject to electrical stress and humidity. Surface impurities may cause a thin film of moisture on the surface of the insulating material to conduct. The subsequent flow of current damages and carbonises the surface, causing a permanent leakage path across the insulator (often called 'tracking'). Such a failure may be accelerated if a surface scratch or crack already exists as a likely path for the discharge.

4.5 Temperature and Failure Rate

In order to relate failure rate to temperature we may consider two
mechanisms of failure. One arises from imperfect sealing or encapsulation,
so that impurities can leak or diffuse into the component so ultimately
causing breakdown. This type of failure will be temperature dependent
since the rate of mass transfer depends upon temperature. As methods
of sealing improve, this kind of failure should diminish. The other model
of failure assumes that some impurity is trapped inside the component
during manufacture. When a mass m of the material has undergone a
chemical change involving the impurity, the component will fail. If the
change occurs at a mass rate of k per hour, the time needed to consume
the mass m is

$$t = \frac{m}{k} \text{ hours}$$

Thus the failure rate for this component is

$$\lambda = \frac{1}{t} = \frac{k}{m} \text{ faults per hour}$$

Amongst a batch of components we expect a variation in the critical
mass m and also a variation in the reaction rate k since this depends
upon several factors, including the amount of material present. For
any particular component, however, the value of m is fixed, and any
factor which changes the value of k will cause a proportional change
in the failure rate λ. This will hold both for a particular component,
and for the average failure rate measured from a batch of components.

An important factor which alters k is the reaction temperature.
There are two expressions used by chemists to relate temperature and
reaction rate. The more exact of these is the Eyring expression

$$k = aT \exp(-b/T) \tag{4.1}$$

where a and b are constants and T is the absolute temperature. This
expression does not lend itself very conveniently to graphical use
since it cannot easily be represented as a straight line or a simple
curve. For some applications the range of temperature involved is
fairly small and the variable T in the first term can be replaced by an
average value without substantial loss of accuracy. Thus equation (4.1)
reduces to

$$k = a \times T_{av} \times \exp(-b/T)$$

Since a and T_{av} are both constants for any particular range of temperature, we may replace their product by a new constant c giving the Arrhenius equation

$$k = c \times \exp(-b/T) \tag{4.2}$$

This was originally derived empirically to fit experimental results, and is much easier to use for graph plotting. If we take logarithms of both sides, we have

$$\log k = \log c - \frac{b}{T}$$

Thus if the component failure rate λ which is proportional to k is plotted on log-linear graph paper against $1/T$, a straight line graph should be obtained whose slope is $-b$. Experimental results usually confirm this, but it is true only if the failure mechanism is the same for all components. This is because the quantity b depends upon the energy changes in the reaction, and if more than one chemical reaction is causing failure, each will have its own different value of b, and the test results will not lie on a single straight line.

If the failure mechanism is known, the value of b can be computed, and so equation (4.2) can be used to extrapolate down to working temperature from results taken at high temperatures. This is the basis for accelerated life testing where in order to shorten the test period, components are tested at temperatures much in excess of normal working temperature.

The form of equation (4.2) means that the failure rate of a component should increase by a fixed ratio for a given temperature rise. For reliability studies it is convenient to take a temperature increment of 10°C. The corresponding factor for the increase in failure rate depends upon the type of component, but is generally about 2. For silicon transistors with plastic encapsulation, a factor of 2·1 for the increase in failure rate for a 10°C rise in junction temperature is suggested. This is derived from failure rate information obtained from life testing. Figure 4.2 shows a typical curve for plastic-encapsulated transistors, based upon the failure rate with a junction temperature of 25°C. The ordinate gives the ratio of the failure rate at any other temperature to the failure rate at 25°C.

The horizontal axis has linear gradations for the variable $1\,000/T$, but the corresponding temperature scale calculated from this is, of course, non-linear.

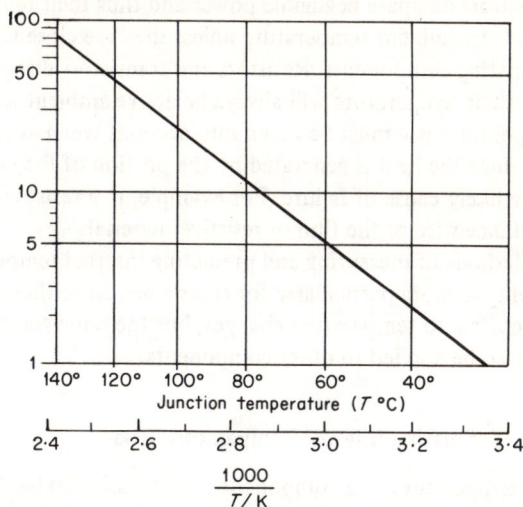

Fig. 4.2 Typical relation between temperature and failure rate for silicon transistor

Other types of component will have somewhat different slopes for the failure-rate temperature curve, for example, a lower factor of 1·25 is quoted by one manufacturer for MOS integrated circuits. These have a failure mechanism different from that of bipolar transistors, with a lower activation energy. A ratio of 2 is suitable for plastic film capacitors, and a somewhat smaller value around 1·5 for resistors. Most chemical reactions which proceed at a significant rate over the temperature range experienced by electronic components have an acceleration factor of about 2–3 for a 10°C rise. Consequently, component failures which involve a chemical reaction generally have a similar temperature dependence.

However, not all components show this simple failure pattern in which the failure rate increases by a constant factor for every 10°C temperature rise. Paper capacitors and mica capacitors generally have a more complicated relation between failure rate and temperature in which the acceleration factor for a 10°C rise increases as the temperature exceeds 60°–80°C, and is also affected by the electrical stress.

An important point to consider when assessing the effect of temperature on failure rate is the degree of internal heating which the component experiences. Low-loss components such as plastic film capacitors dissipate negligible power and thus their temperature corresponds to ambient temperature unless they are close to other heat-dissipating components. Resistors and transistors dissipate heat and thus their temperature will always be above ambient temperature. This temperature rise must be taken into account when assessing failure rate since the heat is generated by the portion of the component which is the likely cause of failure. For example, in a resistor this is the resistance wire or the film of resistive material.

Methods of measuring and predicting internal temperature have been developed more particularly for transistors, since these devices are so susceptible to temperature changes, but the same general techniques can also be applied to other components.

4.6   Estimating Internal Temperature Rise

The temperature of a component which dissipates heat, such as a transistor or resistor can be estimated by using the analogy between heat flow and current flow. A cell which generates current I, and is connected through a resistor R to a current sink at potential V_1 must have a potential

$$V_2 = V_1 + IR$$

In the same way a transistor junction at temperature T_2, having a thermal resistance θ to its surroundings at ambient temperature T_A will be able to dissipate a power W watts where

$$T_2 = T_A + \theta \times W$$

Here the usual units are °C for temperature and °C per watt for θ. The resistance θ can in practice be divided into two components, one internal corresponding to the thermal resistance θ_I between the junction and the transistor case or cooling tab. This resistance is determined when the transistor is made, and cannot be modified by the user. The second component is the thermal resistance θ_E between the transistor case and its surroundings. This can be reduced by using heat sinks, cooling clips, or increasing the rate of flow of air over the transistor by forced ventilation. The full expression thus becomes

$$T_2 = T_A + (\theta_I + \theta_E) \times W \tag{4.3}$$

This expression is widely used for transistor circuit design since the temperature T_2 of the base—collector junction is the temperature which determines the transistor's performance. θ_I is generally given in the transistor specification, together with values of θ_E for various types of mounting and heat sink.

Equation (4.3) is often used for reliability studies to determine whether transistors should be de-rated, that is, used below their specified power dissipation. Thus if we are using a transistor with a value of $\theta_I = 35°C/watt$, a dissipation of 300 mW and an ambient temperature of 50°C, the junction temperature without additional cooling can be estimated from equation (4.2) by using the appropriate value of $\theta_E = 185°C/watt$.

This gives

$$T_2 = 50 + (185 + 35)\,0.3 = 121°C$$

If in the interest of reliability we wish to keep the junction temperature at 100°C or less, we can either modify the circuit to reduce the power dissipation or reduce θ_E by using a heat sink or forced cooling.

If the dissipation is changed, rather than the cooling conditions, the maximum permissible value is given by

$$100 = 50 + (185 + 35) \times W$$

whence

$$W = \frac{50}{220}$$

$$= 0.227 \text{ watts}$$

The transistor must thus be de-rated from 0.3 to 0.227 watts.

Alternatively, if it is preferable to retain the dissipation of 0.3 watts, the external resistance from case to ambient temperature must be reduced to θ_1 where

$$100 = 50 + (35 + \theta_1)\,0.3$$

whence

$$50 = (35 + \theta_1)\,0.3$$

giving

$$\theta_1 = 131.5°C \text{ per watt}$$

Thus the transistor must be given better cooling by using a heat sink to reduce θ_E from 185 to 131.5°C per watt, or must have some forced air circulation.

The above calculation is based upon a constant dissipation in the transistor. If the transistor is pulse driven, the dissipation may jump between two different values corresponding to the two levels of the input signal. The junction temperature then depends upon the thermal time constant of the transistor as well as the steady-state thermal resistances. If the pulses are much shorter than the thermal time constant the junction temperature will depend only upon the mean transistor dissipation. Otherwise the transient changes in junction temperature must be investigated.

Although data for calculating internal component temperature is most readily available for transistors, the same kind of analysis can be applied to resistors, and an expression of the same form as equation (4.3) can be used to determine the temperature of the resistance element.

4.7   The Effect of Voltage Stress

The effect of changing the operating voltage of a component on its failure rate depends upon the type of component. At one extreme is the fixed resistor for which the working voltage has no effect on the failure rate other than that arising from any consequent change in working temperature. Thus for example, if a 5 kΩ ¼ watt resistor has its voltage drop changed from 5 to 10 V, the dissipation will change from 5 to 20 mW. Both of these are so far below the maximum rating of 250 mW that we would not expect any change in the component failure rate.

At the other extreme are non-polarised capacitors using various dielectrics. In all of these any change in working voltage in the region of the rated voltage generally changes the failure rate considerably. An empirical rule which fits most results is the fifth power law, namely that the failure rate is proportional to the fifth power of the capacitor voltage. Where the voltage contains an alternating component the peak value of this is usually taken as the relevant quantity.

If, for example, accelerated tests at 300 V d.c. on a capacitor rated at 200 V d.c. give a failure rate of 0·2 per cent per 1 000 hours, we can use the fifth power law to estimate the expected failure rate in service if the working stress is known. If the capacitor is subject to a voltage whose r.m.s. a.c. component is 50 V, together with a d.c. component of 20 V, the peak voltage is

$$V = 20 + 50\sqrt{2} = 90.7 \text{ V}$$

The ratio $\dfrac{\text{peak working voltage}}{\text{test voltage}}$ is $\dfrac{90 \cdot 7}{300}$ = 0·302.

The failure rate expected in service is thus

$$\lambda = 0 \cdot 2 \times (0 \cdot 302)^5$$

$$= 0 \cdot 0005 \text{ per cent per } 1\,000 \text{ hours}$$

This assumes that the working temperature is equal to the test temperature, otherwise a further adjustment must be made to λ to allow for the temperature change.

4.8 Environmental Factors

We have considered the effect of temperature on component failure rates in previous sections. The other aspects of the environment such as vibration, shock, humidity, and cyclic use of the equipment are often considered together by assigning an environmental factor K_E to each environment. This is used as a multiplying factor to predict the failure rate for various situations. K_E is given a base value of 1 for the most favourable environment, such as fixed digital computer installations, electronic telephone exchanges, underwater repeaters. Satellites in orbit experience no shock or vibrations, but may encounter temperature fluctuations and factors of 1 to 1·5 are used for these.

Equipment which is portable or mounted in road vehicles will encounter some shock and vibration and more rapid temperature changes than fixed equipment, and will be used intermittently. It is consequently given a larger factor in the range 5—10. Equipment on board ship has a similar environment, and the additional possibility of a damp salt-laden atmosphere; the appropriate factor is in the range 10—20.

Airborne electronic equipment has to endure more severe shock and vibration, and will have a wider range of ambient temperature and much more rapid fluctuations of temperature. Factors in the range 30—100 are used, the lower values being typical of a civil aircraft environment, and the higher values more appropriate to military aircraft. The most severe environment in which electronic equipment is used is probably in missiles, for which factors of up to 1 000 have been suggested.

These environmental factors are generally applied to the performance of complete units of equipment and are based upon test and service experience. They represent the effect of the environment on the equipment reasonably accurately, but are much less accurate when applied to individual components. Solid-state devices such as diodes and transistors are fairly resistant to vibration, and thus their failure rates will be much less affected by shock and vibration than will those of electromechanical devices such as relays and potentiometers.

4.9 The Problem of Measuring Failure Rates

In an ideal situation an equipment designer would be given information about component failure rates under various operating conditions together with other technical details in manufacturers' catalogues. Unfortunately the practical constraints of time and expense do not permit the manufacturer to test all of his products under a range of conditions. Modern components are so inherently reliable that even testing one type of component under one set of conditions is an expensive task. For example, a silicon planar transistor in a laboratory or office environment should have a failure rate below 0·02 per cent per 1000 hours. This means that one failure is expected in 5 000 000 component-hours. If 500 transistors can be tested simultaneously, it will take 10 000 hours or almost a year and two months of continuous testing to produce even one expected failure. With such a small number of failures the actual failure rate cannot be estimated very accurately. Furthermore, in order to ensure the quality of each batch of transistors, it would be necessary to hold up delivery of the entire batch for over a year until the result of the life test was known. In view of the wide range of components now manufactured, it is impossible to test each type exhaustively, and a mixture of techniques is usually employed.

Firstly, a small number of components are kept on life test at maximum rating in order to gain some information about performance under rated conditions. This test takes a long time to yield results, so that for a quicker test of consistency from batch to batch, accelerated life tests, usually at high temperature, are conducted. The results of these can be extrapolated down to normal temperatures to yield some additional data.

The acquisition of failure rate data is much more difficult where very high reliability is essential, for example, in underwater telephone repeaters, or communications satellites. For these systems component failure rates of at least an order of magnitude smaller are required, and the testing time to produce even one fault would consequently be about ten times larger. Clearly, for these circumstances conventional life testing is impractical since the entire design of the system and probably the manufacturing methods also would be obsolete before the components could be fully tested.

The amount of information obtained from life testing components is roughly proportional to the number of failures encountered. We have seen that under normal working conditions failure rates are low, and many millions of component-hours of testing are required to obtain useful information. A major source of such data is the fault records of large electronic systems such as digital computers or electronic telephone exchanges. These installations are usually operated almost continuously and have well-controlled arrangements for reporting and clearing faults. One large data-processing installation logged a total of 161×10^9 component-hours during several years of operation and produced much useful information. As in nearly all such cases, however, the data, when finally reported, related to components at least five years old and becoming obsolete.

4.10 Sources of Failure Rate Data

As described in section 3.4, the basic information upon which all reliability assessments are based is a table of failure rates for each class of component used. Furthermore the final figure for reliability or MTBF cannot be more accurate than the set of failure rates upon which it is based. In consequence a major task in any reliability assessment is the acquisition of appropriate failure rate data.

In ideal circumstances such data would be acquired from the serivce records of previous similar equipment, used in a similar environment. In practice, if data were so restricted any project calling for reliability guarantees could not incorporate any significant degree of novelty either in the environment or in the types of component used. In general each new model almost invariably involves some advance in application or technology and so field experience of component failures is not available directly.

In order to ensure that all manufacturers tendering to supply equipment used similar data as a basis for reliability calculations, the U.S. government published a book of component failure rates which was based upon a very large collection of life test results and service records. The current version of this is the MIL-HDBK-217B. Although this relates directly to American components and devices, it is widely

used outside the United States in assessing avionic and defence
equipment as the environment and types of component used are
generally similar. It has the additional advantage that the failure rates
given are based upon a large volume of data.

The tables in the handbook give a set of 'generic' failure rates, for a
set of 11 different environments. These range from 'ground benign', in
which components are subjected to very little environmental stress and
are given optimum engineering operation and maintenance, to the other
extreme of 'missile launch'. For the latter the equipment will suffer
severe conditions of noise, vibration and temperature during missile
launching, boosting into orbit, re-entry to the Earth's atmosphere and
landing by parachute. This environment is also applicable to equipment
on the launching pad which is situated near main rocket engines.

The most severe environment likely to be encountered by civilian
equipment is classed as 'airborne, uninhabited, transport'. This is for
equipment used in the equipment bay, tail or wing of a transport aircraft
where extreme pressure, vibration, and temperature cycling may be
aggravated by contamination from oil, hydraulic fluid or engine exhaust.

The generic failure rates given vary considerably with the environment, the factor relating missile launch failure rate to ground benign
failure rate being of the order of 50 to 150 for most components. Two
procedures are described for deriving the failure rates of components —
parts count reliability prediction and the *parts stress analysis method*.

The first method is intended for use in the preliminary design stages
and requires only a knowledge of the number and quality level of the
components used and their environment. The equipment failure rate is
then the sum

$$\lambda = \sum N_i (\lambda_G \pi_Q)_i$$

where N_i = number of components of ith type, λ_G = generic failure rate
of this type for the particular environment concerned, π_Q = quality
factor for this type of component. The quality factor depends upon the
level of testing, burn-in and inspection to which the component has
been subjected, and perhaps the form of protection or encapsulation.
Thus for severe environments semiconductor devices hermetically
sealed in a ceramic flatpack would be assigned a lower failure rate and
so a lower quality factor than a commercial unit moulded in a plastics
package. Note that the better the quality and the more rigorous the

procurement specification, the lower the failure rate and the lower the quality factor.

This 'parts count' method is adequate for most reliability calculation but at a late design stage it is possible to be more specific about the power dissipation, and so the likely working temperature of each component, the voltage to which it will be subjected and the type of circuit in which it will be used. These factors determine the rate at which it is likely to deteriorate, and the change in characteristics which can be tolerated before the equipment fails to perform correctly.

The failure rate of each component can then be obtained from a failure model which is typically of the form $\lambda = \lambda_B \times \pi_E \times \pi_A \times \pi_Q \times$ etc. The number of factors depend upon the type of component. Here λ_B is the base failure rate, read from a set of curves relating failure rate to temperature. Each curve relates to a particular stress level; for capacitors, for example, stress level is usually taken as the ratio of working voltage to maximum rated voltage. The same stress level factor is applicable for semiconductors, but the important temperature for these components is that of the junction.

The factor π_E depends upon the environment, and is given for the same set of conditions as specified in the parts count method. The degree of drift in characteristics which the component can suffer without circuit failure depends upon the application. Generally device parameters are most important in high frequency applications, less so for linear amplifiers and least for logic circuits. Thus for discrete transistors the application factor π_A is given as

$$\pi_A = 0{\cdot}7 \text{ for logic circuits}$$

$$= 1{\cdot}5 \text{ for linear circuits}$$

$$= 5{\cdot}0 \text{ for h.f. (frequency} > 400 \text{ MHz) applications}$$

As before, the quality factor π_Q depends upon the level of testing and inspection used in manufacture, the type of encapsulation, and the specification to which the devices were purchased. An additional factor for transistors is π_e which depends upon the power rating, and varies from 1 for ratings up to 1 watt to 5 for ratings between 50 and 200 watts.

An additional factor used with integrated circuits is the complexity factor π_C, which increases as the number of gates on the chip increases. Typically the failure rate of a complex 100 to 500 gate package is predicted as 10 to 20 times greater than that of a small scale circuit comprising 1 to 20 gates.

A final factor is used to account for the 'infant mortality' phase when a new design is introduced, or an existing production line restarted after a break of some months. For an initial period of up to 6 months the failure rate calculated using tables and the above factors is multiplied by 10 – called the 'learning factor' π_L. In all other circumstances $\pi_L = 1$.

The use of parts stress analysis is clearly a somewhat formidable task, but it may not be necessary to apply it to every component in the equipment. A parts count calculation will indicate which components contribute most to the overall equipment failure rate. It is clearly worth applying the more rigorous stress analysis to these components, in order to refine the MTBF calculation, but it is unlikely to be worth doing so for those components which add only a small proportion to the overall failure rate.

Since MIL-HDBK-217B was prepared additional data has been collected. Some of this is given in the books by Nicholls (MDR-12) and Klein (MDR-13) which also contain some comparisons between the observed failure rates of electronic equipment and those calculated according to the methods given in the handbook.

Reliability data on electronic components is collected in the United Kingdom, but inevitably on a smaller scale than is possible in the United States. The two main sources are the National Centre of Systems Reliability, Warrington, and R.S.R.E., Malvern.

4.11   Confidence Limits and Confidence Level

Having acquired some information about component reliability from equipment fault records or life testing we encounter a further problem in deducing what figures we should use for design purposes. The difficulty is common to all sampling methods, and arises because we wish to make a statement about the properties of a batch of components when we have tested only a small sample from the batch. The essential fact is that the sample may or may not be typical of the batch. We cannot make any certain statement about the relation between the characteristics of the batch and the sample, we can only make statements which have a certain probability of truth. Thus we may deduce from tests that there is a 90 per cent probability that the failure rate of a certain batch of components under certain operating conditions lies between 0·002 and 0·018 per cent per 1000 hours. There are two items of information in this statement, firstly the numerical limits between which we assert that the batch failure rate lies. This is called the 'confidence interval'. In the example we have a two-sided interval since we specify both an upper and a lower limit. Frequently, in reliability studies we are concerned with a one-sided interval in which we express only one limit.

The second item of information in our statement is the degree of confidence we attach to the assertion. This is called the 'confidence level'. We have used a figure of 90 per cent, so that our statement could also be put in the form 'The failure rate of this batch of components lies between 0·002 and 0·018 per cent per 1000 hours with a confidence level of 90 per cent'.

A one-sided interval is used in a statement such as 'The failure rate of this batch of transistors is 0·01 per cent per 1000 hours or less with a confidence level of 90 per cent'. The confidence level is here given as a probability, but we could also use the frequency basis of probability by restating the assertion as 'Nine times out of ten the failure rate will be 0·01 per cent or less'. This means that if we test many different batches of transistors and obtain identical test results, our assertion based on the test results will be correct in nine out of ten cases on average.

4.12 Evaluating Test Results

The final question to be resolved is the relation between test results and the establishment of confidence limits and a confidence level for the batch characteristics. For a given set of results we can take a number of different confidence levels and evaluate the approximate limits for each of them. A widely used confidence level in reliability assessment is 90 per cent which we will use as a typical value.

The original use of confidence limits and levels was in sampling the output of mechanical components, where the distribution of the batch was known, and a sufficient number of samples were tested to allow a good estimate of mean and variance. Unfortunately, in studying the reliability of electronic components, we typically encounter very few failures or none at all. In these circumstances we have insufficient data to estimate the mean and variance and must rely on other methods.

The most suitable statistical tool for analysing a few experimental values of, say, time to failure for a particular type of component is the 'chi-squared' distribution. The quantity chi-squared is defined as the sum of the squares of K independently chosen members from a normally distributed population with zero mean and unit standard deviation. Thus if x is one member of a normal population with a mean of $\bar{x}$ and a standard deviation of σ, the value of chi-squared is

$$\chi^2 = \sum_1^K \left(\frac{x - \bar{x}}{\sigma}\right)^2$$

The variable in the bracket is derived from x so as to give a mean of 0 (by subtracting $\bar{x}$) and a standard deviation of 1 (by dividing by σ). The mean of the chi-squared distribution is N and it is the same as the gamma distribution for

$$\alpha = \left(\frac{N}{2} - 1\right) \quad \text{and} \quad \beta = 2$$

Both the chi-squared and the gamma distribution tend towards the normal distribution as N tends to infinity.

The chi-squared distribution for $2R$ degrees of freedom with a level of significance α is related to the Poisson distribution with an expected number of failures n_f by the relation

$$\chi^2 = 2n_f$$

The chi-squared distribution is applicable to the analysis of component or system life tests by using tabulated values of the distribution for various levels of significance. The restriction is that the procedure applies only if the failure rate is assumed constant.

To illustrate the procedure, we assume that a number of components are put on test, and the times of the first, second and subsequent failures are noted. We then calculate the number of component-hours of testing to the various failure events. At a given confidence level we can then calculate the lower limit of MTBF that the test has demonstrated. Table 4.1 gives some values of the multiplier K used in calculation as a function of the number of failures.

Table 4.1

Failure number	First	Second	Third	Fourth	Fifth
K	2·30	3·89	5·32	6·70	8·00

The multiplier K is equal to the ratio

$$\frac{\text{component-hours to failure}}{\text{demonstrated MTBF at 90 per cent confidence level}}$$

Thus if we wish to demonstrate an MTBF of at least 1 000 hours, we must test for $1\,000 \times 2 \cdot 30 = 2\,300$ component-hours before the first failure occurs, or $3 \cdot 89 \times 1\,000 = 3\,890$ hours before the second failure occurs. Although the reliability of complete equipment is usually specified in terms of its MTBF, component reliability is generally given as a failure rate, the reciprocal of MTBF.

If, for example, we have 1 000 transistors on test for two years without failure, we have a total of $17 \cdot 52 \times 10^6$ component-hours. We take this as the minimum time to the first failure, so $K = 2 \cdot 30$.

Thus the demonstrated MTBF is

$$M = \frac{17 \cdot 52 \times 10^6}{2 \cdot 30}$$

$$= 7 \cdot 6 \times 10^6 \text{ hours}$$

This corresponds to a failure rate of

$$\frac{1}{7 \cdot 6 \times 10^6} = 1 \cdot 31 \times 10^{-7} \text{ per hour}$$

$$= 0 \cdot 0131 \text{ per cent per 1 000 hours}$$

at 90 per cent confidence level.

For systems such as communications satellites or underwater telephone repeaters we require a considerably lower failure rate than this. Consequently, life testing under normal operating conditions is not a very practical proposition for these applications. Even for the lower reliability demonstrated in the above example we must wait two years for the result.

We have assumed above no failures during the test so the total component-hours is simply the product of the number of components on test and the test duration. If, however, we test only a few models of a complete electronic unit, the calculation of the total component-hours is a little different.

We will assume that four models of a new unit are tested, and failures occur at 507, 720, 860 and 1 020 hours. Each unit is withdrawn from test when it fails. What is the demonstrated MTBF at 90 per cent confidence level?

The total number of unit-hours is

$$T = 507 + 720 + 860 + 1\,020$$

$$= 3\,107 \text{ hours to the fourth failure}$$

From the table, $K = 6 \cdot 70$; thus the MTBF demonstrated is

$$M = \frac{3\,107}{6 \cdot 7} = 464 \text{ hours at least}$$

with 90 per cent confidence level.

If the unit is intended to be repairable, it may be a more realistic procedure to repair each failure as it occurs and return the unit to the test bay.

The total number of unit-hours then accumulated is simply the product of the test duration and the number of units.

Taking the above figures, if units were repaired and returned to test the total number of unit-hours would be

$$T = 4 \times 1\,020 = 4\,080$$

less a small time taken to repair the faults.

4.13 Sequential Testing

The test procedures discussed earlier were intended to determine component or system reliability as accurately as possible in the time available. When a buyer stipulates a minimum standard of reliability, for example, by specifying a minimum MTBF for a particular electronic unit, a different test criterion is involved. A sample of the production is tested and for economy it is desirable to do the minimum amount of testing necessary to establish with a prescribed confidence level whether the specified MTBF has been achieved.

Thus the test schedule is designed to determine quickly whether the achieved reliability is much better or much worse than that specified. Only if the achieved MTBF is near to the specified MTBF (M_s), is an extended test required.

The usual procedure in this case is to use sequential testing. A graph is drawn of the progress of testing, plotting the number of failures vertically, and the accumulated unit-hours horizontally. The area of the graph is divided into three zones as shown in figure 4.3, corresponding to a decision of accept, reject, or continue testing. It is customary to insist on a minimum test period of, say, $4M_s$ unit-hours even if no failure has occurred. In order to provide an upper limit to the testing required, if the actual MTBF is close to M_s the test is usually truncated. This means that if the graph is found to lie within the central band, after a prescribed amount of testing, typically about $30\,M_s$ unit-hours, testing

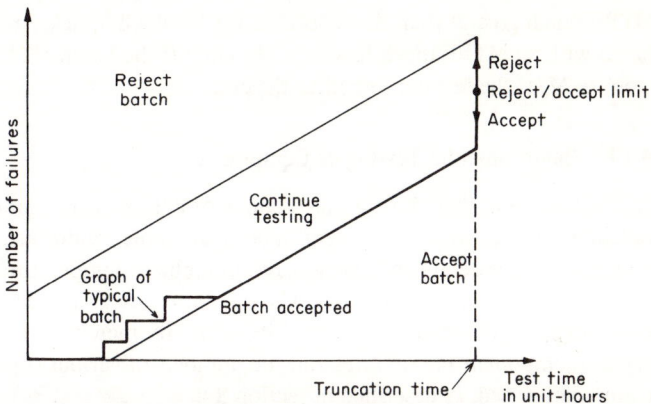

Fig. 4.3 Plot of sequential test

is concluded, and if less than a certain number of failures has occurred the batch is accepted. This sequential test procedure was originally proposed by the American Advisory Group in Reliability of Electronic Equipment (AGREE) and has since been adopted for validating the reliability of much military equipment and electronic systems for civil aircraft.

The precise positions of the accept and reject lines require prior agreement between the supplier and user. They are usually placed so as to equalise the risk to both parties at, say, the 10 per cent level. This means that there is a 10 per cent chance of a good batch (which meets the MTBF requirement) being rejected — the supplier's risk — and also a 10 per cent chance of a bad batch being accepted (the user's risk). It is customary to use different MTBF figures for the two cases, the user's risk being specified in terms of the minimum MTBF M_1 needed to meet operational requirements, and the supplier's risk in terms of a higher value, typically $M_2 = 1.5\,M_1$.

Thus for testing production equipment under the AGREE procedure, the two risks are 10 per cent and the discrimination ratio M_2/M_1 is 1·5. If, for example, the operational requirement calls for a minimum MTBF of 100 hours, the user has a 10 per cent chance that the test will accept a batch of equipment which has MTBF of less than 100 hours, and the supplier takes a 10 per cent risk that a batch with an MTBF of over 150 hours will be rejected. The numerical values of risk and discrimination ratio are a compromise between the need for a reasonably effective test and the time and cost of conducting it. The sequential test procedure outlined above will quickly accept a batch with an MTBF much greater than that specified (M_s) and will quickly reject a batch with an MTBF much less than M_s. Only if the batch MTBF is close to M_s is the test continued to the end.

4.14 Environmental Testing of Equipment

As we have seen, there is an ample volume of data available about the reliability of electronic equipment under favourable conditions, with no vibration, moderate and fairly constant ambient temperature, and clean and fairly dry surroundings. The purchaser of equipment for such conditions can obtain an estimate of its reliability merely by using typical component failure rates and the numbers of various types of components used, as described in section 3.4.

The purchaser of equipment to be used in an arduous environment is, however, in a much more difficult position. Although some information is available about failure rates at elevated temperatures, when changing rapidly between extremes of temperature, when in a very damp environment, or subject to vibration, etc., it is most unlikely that the particular combination of all these different conditions which the equipment must withstand has been used for testing. Also since comparatively little of the total quantity of electronic apparatus produced is used under extreme conditions there is not much reliability data available from use in the field.

A further difficulty is that under severe conditions component reliability is much dependent upon the way the component is mounted, supported, cooled, or protected from the worst effects of its environment. Consequently the same component in the same environment may have a wide range of failure rates depending upon the way it is mounted, cooled, etc.

For the above reasons the purchaser of equipment for use in severe environments cannot estimate its probable reliability with any degree of confidence by, for example, counting the components on the circuit diagram, and assigning to each type a failure rate. Thus if reliability is important to the user, he will require some demonstration that the equipment meets its specification (usually a minimum value of MTBF).

This requires that the equipment should be put on life test, in a test chamber which as far as possible duplicates the environmental conditions which the equipment must withstand in service. This is called 'environmental testing', and calls for a complicated and expensive test facility, with provision for controlled temperature and temperature cycling between $-65°C$ and $+70°C$, typically, controlled humidity, atmosphere, water and salt water spray, and variable vibration along several axes. In addition to providing the full working environment, the test rig must also be able to test the electrical performance of the equipment to its full specification.

In order to cater for a range of environments, the AGREE committee proposed four different degrees of environmental stress. In the first two the temperature is varied by only $±5°C$, but the third and fourth categories call for temperature cycling between wide limits of heat and cold. In order to simulate the temperature change and the rate at which it occurs, powerful heating and cooling equipment is needed, particularly for testing airborne apparatus. A fast jet aircraft which begins a flight from an airfield in the tropics may in a short time be flying at high altitude in temperatures below freezing, so the environmental test chamber must able to match this rapid change in temperature.

The time during which the temperature is kept constant depends upon the time the equipment under test takes to attain a steady temperature. The time for a hot or cold part of the cycle is thus adjusted to ensure that virtually steady-state temperature conditions have been reached before moving to the next part of the cycle. A period of three to five times the thermal time constant of the equipment is a reasonable estimate, with a minimum of 3 hours. At least four temperature cycles each 24 hours is suggested.

The equipment is vibrated by a moving-coil unit fed from a high-power audio amplifier, or a mechanical vibrator. In order to ensure that all resonances which may exist are examined, the amplifer input may be either a gliding frequency, moving up and down between specific limiting frequencies, or a random 'white noise' signal, filtered through the appropriate band-pass filter. For type-testing in order to prove the design it may be necessary to submit the equipment to several vibrational tests along different axes. Also some users may call for an impact or shock test, in addition to continuous vibration. Equipment which may be subjected to a dirty or fume-laden atmosphere also needs testing in a similar atmosphere which contains appropriate contaminating gases such as sulphur dioxide.

Table 4.2
Test levels proposed by AGREE Committee

Conditions	1	2	3	4	5
Temperature	25° ± 5°C	40° ± 5°C	−54° to +55°C	−65 to +71°C	50° ± 5°C
Vibration	None	0·01 in peak-to-peak at 25 ± 5 Hz	As 2	As 2	0·01 in peak-to-peak at 20 to 60 Hz
On-off cycling	3 hours on + time to stabilise temperature	As 1	As 1	As 1	As 1
Input voltage	Nominal	Maximum specified +0 −2 per cent	As 2	As 2	Nominal

Table 4.2 shows the five main categories of test schedule proposed by the AGREE Committee.

A further variable during the test may be the power supply. This normally has a tolerance stated in the equipment specification, and the test may call for full performance testing at both extremes of the permitted variation, or only at, say, maximum voltage. In addition, if the equipment is subject to intermittent use, a duty cycle may be specified in which the equipment is switched on and off at fixed intervals of time.

Although it may appear an easy task to measure the characteristics of a particular environment and simulate them in the laboratory, great care is needed to ensure that all factors in the environment, even though they appear irrelevant, are considered. The need for care is exemplified by the cautionary tale of an American radio aerial array and its corrosion problems. This used light alloys and was designed for small naval vessels. Despite a full type test in dry, damp and salt water spray conditions it unexpectedly began to corrode after a short time in service. After considerable investigation it was discovered that the missing ingredient in the environment test was seagulls! Although the aerial could withstand salt water spray alone, salt water together with seagull droppings soon caused corrosion.

Similar unusual corrosion problems may affect equipment used in chemical plant, but it is usually possible to seal the equipment from the corrosive fumes, and if necessary pipe a supply of clean air in for cooling.

4.15   Screening and Failure Mode Analysis

We have noted in section 4.9 that modern components under mild ambient conditions are very reliable and consequently life tests to determine their failure rate are lengthy and expensive. It has thus been argued that, since the results from these tests are only available perhaps years after the components were made, it is more profitable to devote the effort not to testing but to determining the cause of the failures which do occur. If the investigation shows any deficiency in the production process, this can be remedied with the expectation of improved reliability.

Such measures are usually more profitable in the early stages of production, for example, in the early days of silicon transistors an unsuitable method of connecting the chip to the leads of the encapsulation caused metallurgical failures such as 'purple plague'. As more is learnt about making components the gross errors are eliminated, and ultimately the majority of failures are caused by deviations from the specified process rather than inherent weaknesses in the process itself.

There is still the difficulty that this failure analysis technique requires failures to operate upon. Some of these may come from equipments in the field, but the manufacturer can generally rely for more accurate information on their operating conditions and environment if these are under his own control. Frequently failures are accelerated by operating components at elevated temperatures or over their voltage or current ratings. The operating temperature cannot, however, be raised too much, otherwise new failure mechanisms appear which are not significant at normal temperatures. For example, soldered joints may begin to fail at excessive temperatures whereas they may be completely reliable at normal temperatures.

Thus it is important to establish that the failure mode which occurs in accelerated life tests is the same mode as occurs in normal working conditions. Some check on this can be made by plotting test data as a graph of $\log \lambda$ against $1/T$, as in figure 4.2. If the graph is a continuous straight line, the same failure mechanism is likely to be operative at all temperatures. If, however, there is a change of slope at some point, this denotes a change in activation energy and so a change in failure mechanism. In the first situation, it is usually safe to extrapolate high temperature test results down to normal working temperatures.

4.16　Screening Tests

In applications which require high reliability, one method of eliminating components which are liable to fail during their working lives is to discard those which show a significant drift in characteristics. Semiconductors are particularly suited to this procedure.

The characteristics of all components are measured initially, and then they are put on life test. After an appreciable period of test the characteristics are again measured and any components which show a significant change are rejected. There is some evidence that a number of components which would not fail in service are rejected, but the majority of potential failures are eliminated.

This process is called 'screening'.

The demands of production schedules and the cost of testing usually permit only a short test generally less than 1 000 hours, and relatively few components, say batches of hundreds or thousands, are involved. In order to eliminate potential failures in this time interval it is necessary to subject the components to some degree of overstress, the most generally useful one being temperature. A variety of combinations of time and temperature have been proposed for screening. Some integrated circuits are given a hot cycle of 250 hours at 125°C. Another plan which fits conveniently into production schedules is a week (168 hours) at 125°C. However, for American telephone equipment the latter screen was found rather ineffective, and was replaced by a cycle of 16 hours at 300°C. This gives a much greater probability of detecting a potential failure, and is equivalent to about 100 000 hours at 125°C.

A similar screen is used by the British Post Office (BPO) for ultra-reliable transistors for underwater repeaters; this is a little more severe as it requires 24 hours' treatment at 310°C.

The device characteristics, including leakage currents for semiconductors, are measured before and after the test and any component showing a substantial change in characteristics is rejected. This screening procedure appears to be fairly effective although it causes a number of components to be rejected which may not have failed subsequently in service.

Although a high temperature test is the most widely applicable one, other tests have also been used to detect different modes of failure, namely

    (a) A short vibration test, with electrical characteristics being monitored. This detects intermittent faults, leads nearly touching, or loose metallic particles trapped inside a can.

    (b) Constant acceleration (typically 20 000 g in a centrifuge). This checks the bonding of leads in a semiconductor device.

    (c) Insulation between leads and case.

    (d) Test for leaks in a hermetically sealed component.

    (e) Where a resistor is liable to fail due to oxidation of the resistive material, the load test at high temperature can be made more onerous by using an atmosphere of oxygen.

This screening procedure is clearly an expensive procedure and its cost can be justified only where extreme reliability is demanded. If an underwater repeater fails in a transatlantic telephone cable during a period of bad weather, the cost of diverting a cable ship to recover the faulty repeater and replace it, together with the loss of revenue while the circuits are inoperative may easily exceed a million pounds. Consequently it is economically worth while to use the most reliable components available, even at a considerable increase in cost.

4.17   The Demand for Extreme Reliability

In this connection some typical requirements indicate the degree of reliability now called for in some high-reliability systems. For example, the BPO target for transistors used in underwater repeaters is one failure in 1 000 transistors in twenty years. The complete system uses many repeaters, with some degree of redundancy, and has a design life of twenty years. The transistor failure rate demanded is 0·00057 per cent per 1 000 hours, or 5·7 per 10^9 hours. An early lunar probe, Mariner 2, was required to have a 95 per cent probability of success in accomplishing a four month mission. It contained about 3 000 components an consequently required the average failure rate for all components to about 5·7 per 10^9 hours. Since the passive components are generally more reliable than transistors and diodes, this figure allows a somewhat larger failure rate for transistors than the BPO requirement.

    A similar demand for extreme reliability occurs in electronic telephone exchanges, for which a typical reliability requirement is a downtime of only a few hours in a twenty year life. Such a specification necessitates a degree of redundancy, but still requires extremely reliable components.

The maintenance records of early American electronic telephone exchanges show component replacement rates as indicated in column A of table 4.3. For comparison the table also shows in column B the failure rates proposed by the BPO for calculating the reliability of future British electronic exchanges. The BPO figures are somewhat greater than the American figures, partly because they relate to the use of professional grade components with a considerable degree of de-rating, whereas the American exchanges use some special high-reliability components.

Table 4.3
Failures per 10^9 hours

Component	Application		
	A	B	C
Carbon and metal film resistors	1	8	0·5
Small fixed capacitors	5−10	18	1
Low power silicon transistors	10	45	20
Sealed relays	100	−	−
Integrated circuits − small scale	−	80	15
Reed relays	−	23	350
Switching diodes	−	16	2
Soldered joint	−	1·8	0·5

The environment of electronic telephone exchanges is generally a favourable one, so that the figures given above represent the lowest failure rates which can normally be expected from the particular types of component used.

Telephone equipment is generally expected to have a working life of 20−25 years, and consequently long-term reliability is a major consideration in the design of any electronic telephone equipment. Much electronic equipment, however, has a working life of only 5−10 years before being replaced due to obsolescence or changed requirements. For this class of equipment a rather more economical design is appropriate, with less de-rating of components.

Even so, when used in a benign environment such equipment can have a surprisingly high reliability. The first electronic digital computer built failed initially about every $1\frac{1}{2}$ hours. A current minicomputer of much greater power would have a failure rate nearer to 1 fault per year.

Some extracts from the field service data of a large U.K. computer manufacturer are shown in column C of table 4.3. These were collected over a five-year period and relate to digital circuits with carefully regulated power supplies, adequately cooled and with a temperate environment. The diodes and transistors were used in low power logic circuits. The soldered joints were produced mechanically by flow soldering; the failure rate for hand soldered joints was 2·0 per 10^9 hours. The capacitors were plastics film capacitors of a particular type; the average for all kinds of plastics film capacitors was 3 failures per 10^9 hours.

The combined failure rate for carbon film and metal film resistors shows them to be extremely reliable. An interesting comparison with the figure given is provided by a recent analysis of the last 14 years of operation of submerged telephone repeaters incorporating metal film resistors. These work in a very benign environment with little change of temperature and an atmosphere of dry nitrogen. During the period 9×10^9 component hours of operation were accumulated without a single fault. Assuming that a fault occurred immediately afterwards, the observed failure rate was 0.11 per 10^9 hours.

The performance of the reed relays in the computer application appears relatively poor since recent service records of the BPO TXE2 exchanges show an observed failure rate of just below 15 per 10^9 hours.

Having discussed in some detail the failure rates of components and the factors which affect them, in the next chapter we examine the task of the designer in selecting appropriate components, deciding on their values and other characteristics, and assembling them into a fully reliable system.

PROBLEMS

1. Life tests on a type of capacitor gave a failure rate of 3 per cent per 1 000 hours when tested at 80°C and 100 V d.c. What would be the approximate failure rate at 45°C with a potential of 15 V a.c. super-imposed on 30 V d.c.? Assume that the failure rate is doubled for a 10°C temperature rise.   [0·0093 per cent/1 000 hours]

2. A batch of transistors tested at a junction temperature of 120°C gave a a failure rate of 0·12 per cent per 1 000 hours. If the internal thermal

resistance from junction to case is $40°C$ per watt, and the heat sink resistance is $160°C$ per watt, calculate the maximum permissible transistor dissipation which can be permitted if the failure rate is not to exceed 0·02 per cent per 1000 hours. Assume that the failure rate increases by a factor of 2·1 for each $10°C$ increase in junction temperature. The ambient temperature in service is $45°C$.   [254 mW]

3. Three sets of airborne communications equipment were tested in an environmental chamber and failed after 460, 570, and 810 hours respectively. What is the demonstrated MTBF at 90 per cent confidence level if they are assumed to have a constant failure rate?   [346 hours]

4. A digital computer containing 2 500 integrated circuit packages is used as a test-bed to assess the reliability of the packages. For how long must the computer continue to operate without a package failure in order to demonstrate a package failure rate of 0·02 per cent per 1000 hours at a confidence level of 90 per cent?   [4 600 hours]

5 Designing for Reliability

5.1 Aspects of Reliability

In the design of electronic equipment, reliability is one of many requirements, such as weight, size, cost, power input and performance which must be given adequate attention in arriving at the most effective product. Like most other characteristics, reliability is a feature which must be given due attention at all phases of the design, and cannot satisfactorily be added on after the prototype stage has been reached. It is, however, useful to consider the problem of designing reliable equipment under four headings, system design, circuit design, choice of components and mechanical design, including packaging and assembly methods.

Under system design we include the subdivision of the system into suitable sub-assemblies, the provision of test facilities, alarm and failure indication, and the use of some redundancy if needed to meet reliability requirements. Many of these factors will be largely decided by the particular requirements of the system and the environment in which it must work so few general principles can be distinguished. We have investigated the improvement in reliability which redundancy can provide in chapter 3; some of the practical problems of introducing it are considered later in sections 5.11 to 5.16.

There is no room in a book on reliability for a full treatment of electronic circuit design, but some fundamental principles involved are important. The first of these is the concept of what is called 'worst-case' design. This is largely concerned with the difference between the theoretical solution to a network problem given the values of all components and e.m.fs in the circuit, and the actual behaviour of a real electronic unit of which the circuit on paper is an ideal and generally approximate model.

5.2   The Effect of Tolerances

Let us consider a very simple example in which we require to know the current which will flow through a $1\,k\Omega$ resistor when connected to a 9 V supply. This is a completely specified network, and the current must be exactly $9/1 = 9\,mA$.

If we change the question to a more practical one, however, we will usually not obtain a unique answer. If we now enquire what current will flow if we take a resistor from the stores bin labelled '$1\,k\Omega$' and connect it across a dry battery labelled '9 V', we must make further enquiries before being able to answer. This is because there is now a degree of uncertainty associated with the two values we are given. The label '$1\,k\Omega$' on the bin indicates only the nominal value of the resistor. It is unlikely to be exactly $1\,k\Omega$ but will be near to that value. The limits between which it must lie are given by the 'tolerance' of the resistor. The maximum standard tolerance for a carbon composition resistor is 20 per cent. If we have such a resistor it should have a value between $1\,k\Omega - 20$ per cent and $1\,k\Omega + 20$ per cent, that is, between 800 and $1\,200\,\Omega$.

If we have picked a carbon film resistor it is more likely to have a tolerance of 5 per cent and should thus lie between 950 and $1\,050\,\Omega$.

Similarly the battery will probably not have a terminal voltage of exactly 9 V. If it is new and of ample capacity, so that the load causes negligible voltage drop the initial terminal voltage may be as much as 9·5 V. On the other hand the voltage may drop as low as 7·5 V before the battery is regarded as exhausted.

We can thus calculate the maximum possible current by taking the maximum terminal voltage in conjunction with the minimum value of resistance. This gives $10·5/0·8 = 13·1\,mA$ (assuming a 20 per cent resistor tolerance). At the other extreme the minimum possible current occurs with minimum battery voltage and maximum resistance and will be $I = 7·5/1·2 = 6·25\,mA$.

There is thus nearly a 2:1 variation between maximum and minimum currents. If we are concerned with designing for a specified level of current, for example, if our resistor represented a relay with a resistance of $1\,k\Omega \pm 20$ per cent, the critical 'worst-case' would be that involving minimum current, namely 6·25 mA. Consequently for reliable operation of the relay we should require one which would be guaranteed to operate on less than, say, 6 mA.

If on the other hand we are investigating the temperature rise of the resistor or relay coil, which depends upon its dissipation, we should look for the worst case involving maximum dissipation, that is, maximum voltage and current.

The general procedure for worst case design is to examine each specified characteristic of a circuit block, such as power output, gain frequency response, etc., with all of the components which have tolerances fixed at their least favourable values. If the circuit will perform to specification for this set of values it should operate satisfactorily with all other values of component within the given tolerances.

5.3   Worst-case Design of Switching Circuit

As a slightly more complicated example we will consider the switching circuit shown in figure 5.1. The resistors in this case have 10 per cent tolerance, and the transistor current gain lies in the range 20–70. The problem is to specify the minimum input voltage V_s which will ensure saturation of the transistor, with an overdrive factor of 2·0. This means that the base current must be at least 2·0 times greater than required just to ensure saturation, so that the switching operation can occur rapidly. The transistor potentials when saturated are $V_{be(s)} = 0.8$ V, $V_{ce(s)} = 0.3$ V.

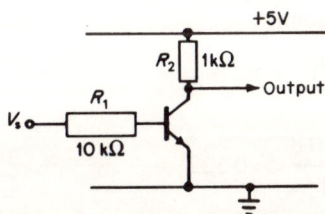

Fig. 5.1 Transistor switching circuit

In this example we assume a constant supply voltage, so that the parameters subject to tolerance are the resistor values and the transistor current gain. For worst-case design we require the situation which will produce the minimum base current for a given value of V_s and will require the maximum base current to meet the saturation criterion.

The component values must thus produce the maximum collector current possible and can be listed as follows

(a) R_1 maximum
(b) R_2 minimum
(c) current gain minimum

The component values to be used for calculation are thus

$$R_1 = 11\,\text{k}\Omega, \quad R_2 = 900\,\Omega, \quad h_{fe} = 20$$

The collector current is

$$I_c = \frac{5 - 0\cdot3}{0\cdot9} = 5\cdot22\,\text{mA}$$

The base current required is thus

$$I_b = 2\cdot0 \times \frac{I_c}{h_{fe}}$$

$$= 2\cdot0 \times \frac{5\cdot22}{20} = 0\cdot522\,\text{mA}$$

The base current available is

$$I = \frac{V_s - 0\cdot8}{11}\,\text{mA}$$

Thus in order to meet the specified conditions for all possible component values we require

$$I > I_b$$

or

$$\frac{V_s - 0\cdot8}{11} > 0\cdot522$$

Thus

$$V_s - 0\cdot8 > 5\cdot74$$

whence

$$V_s > 6\cdot54\,\text{V}$$

Thus the worst-case combination of component values requires a drive voltage V_s of at least 6·54 V. We may thus specify a minimum drive of, say, 7 V and can be certain that any circuit built from components with the nominal values and tolerances given will switch correctly.

The above analysis is concerned only with the conditions required to saturate the transistor and produce a logical '0' signal (minimum voltage level) at the collector. In general the other condition which requires examination is that of a logical '1' output (maximum voltage level) for which the transistor should be cut off, that is, passing only leakage current. If the transistor is a silicon type, the base will not conduct until it is raised to a potential of about 0·6 V. Thus if V_s is, say, 0·4 V or less the transistor will be cut off as required. This condition depends upon the characteristics of the transistor alone and the values of R_1 and R_2 are almost immaterial. Thus if we are using worst-case design methods to decide suitable values of R_1 and R_2 the cut-off condition provides no information and we must rely on the data from the saturation condition.

5.4 The Difficulties of Worst-case Design

Although worst-case design methods are widely used for circuit design the procedure outlined above can be criticised in two major respects. Firstly, using normal manual design we assume that the worst-case value of each component will be either the maximum or minimum value; this is usually a valid assumption for digital circuits but not always for linear circuits. For example, if we are concerned with maximum power dissipation in a transistor, this may occur with maximum supply voltage, minimum collector load resistance, and some intermediate values of the bias resistors rather than the maximum or minimum values.

In order to discover the worst-case configuration in such a circuit, we need to take, say, ten intermediate values of the bias resistors, as well as the maximum and minimum values. If there are two of these resistors there are in all 144 different combinations of their values to be investigated in addition to any other component variations. Clearly, if many component values need this kind of detailed investigation, a great deal of computation is needed, and the assistance of a digital computer is almost essential.

The basic assumption of the worst-case design procedure is that the worst-case combination of component values can fairly easily be deduced. It is then necessary to compute the performance of this case alone in order to ensure that all other combinations of component values will have more than adequate performance.

If, however, we need to investigate many possible combinations in order to ensure that we have discovered either the worst case or a near approximation to it, the economy in computation which is a major attraction of the process no longer applies. The amount of computation may be several orders of magnitude greater than it would be if the worst case were easily found by inspection; in this situation we can often use the computing effort better by adopting a different design procedure.

The other objection to worst-case design is that the average circuit based upon it has a considerably higher performance than the minimum required. The more components in the circuit the greater will be the variation in performance between the worst case and the best case. Since the design selects component values which ensure a satisfactory performance for the worst possible case, the other combinations of component values will have a higher performance than specified and will thus be to some degree uneconomical. The greater the number of components, the greater the spread of performance and the less economical the design will be.

It may well be possible to produce a more economical design by aiming for a lower mean performance, and accepting the consequence that a few of the production units whose component values are near to the worst-case values will be unsatisfactory and rejected at the testing stage. In a linear amplifier, for example, this relaxation of performance may allow the designer to use three stages instead of four, so saving, say, 20 per cent of the component cost. If only 5 per cent of the production is rejected for inadequate performance, the resulting unit will be cheaper and more reliable since it has fewer parts.

5.5 Statistical Design

This process is sometimes called 'statistical design' and it depends upon a knowledge of the distribution of component values between the tolerance limits, and the ability of the designer to assess the proportion of the production which would be rejected as unsatisfactory. Since it would be much too expensive to build, say, 1 000 amplifiers to a given design and measure their performance, it is necessary to simulate the process using a computer. A random value within the tolerance range for each component is calculated using a random number generator, and the performance of the circuit is calculated. The procedure is repeated a thousand times or more, and from the results the designer can assess the likely reject rate for his design.

This procedure requires generally a greater amount of computation than worst-case design and so it is justified only for circuits which are produced in large numbers. Some of the first circuits to which the procedure was applied were the logic circuits for computers, of which many millions are now produced every week.

By contrast the simpler worst-case design process is more suitable for low-volume items and 'one-off' equipment.

5.6 Component Selection

Having decided on the values of the passive components, the designer must decide what type of component to specify. The decision largely depends upon the tolerance which the circuit will permit. For example, if a capacitor is required for an application not involving high ambient temperature the cheapest and smallest type of capacitor is an electrolytic one. But since its tolerance may be −50 per cent, +100 per cent it is quite unacceptable if only a 5 per cent variation from the nominal value is permissible. Again it has high losses when used on an alternating current supply, and requires a polarising voltage. Thus it cannot be used where a low-loss capacitor is required, or where the voltage across it may change sign.

Resistors are available in a wide range of tolerance, varying from ±20 per cent for a standard carbon composition unit to ±0·1 per cent or less for a precision wire-wound resistor.

In addition to the initial or selection tolerance of the component the designer must also make allowance for change in value due to other causes. The important ones are temperature and age.

The temperature coefficient of a carbon composition resistor may be up to 0·1 per cent per °C, so that its value may change by 10 per cent for a temperature variation from, say −30°C to +70°C. This is less than the temperature variation which some military or airborne equipment may experience.

At the other extreme a precision wire-wound resistor which embodies wire drawn from a low temperature coefficient alloy may have a temperature coefficient of only 5 parts per million or 0·0005 per cent per °C.

Carbon composition resistors are generally avoided when designing high reliability equipment in favour of film resistors in which the resistive material is a film of metal, metal oxide or carbon. The metal oxide and metal film types in particular have low temperature coefficients in the region of 20–200 parts per million per °C, and a lower drift in value with time than carbon resistors.

In addition to the variation in value with temperature, resistors exhibit a small change in value when soldered into place.

Capacitors also change in value with temperature, but to a smaller extent; for example, mica and plastic film capacitors have temperature coefficients in the region of 70 and −200 parts per million per °C.

A particular type of small capacitor using a ceramic dielectric can be made with several different values of positive or negative temperature coefficient. By combining these with other types of capacitor it is possible to produce a combined capacitance of any required value with a very low overall temperature coefficient where a constant value capacitor is required. Alternatively, if a stable tuned circuit is required the capacitor can be arranged to have a small negative temperature coefficient to cancel out the small positive temperature coefficient of the associated inductor.

On the other hand some ceramic capacitors which incorporate high permittivity dielectric may change in capacitance by over 40 per cent for a temperature change from 0° to 80°C. Electrolytic capacitors are more stable but may vary by 30 per cent over the same temperature range.

Some temperature effects are unavoidable; for example, the leakage current of semiconductor diodes and transistors doubles for roughly a 9°C temperature rise. This depends upon the physical properties of silicon and occurs in all p–n junctions in this material. This leakage is so great that it precludes the use of germanium transistors at high temperatures, but silicon transistors in which the leakage is two to three orders of magnitude lower are usable at junction temperatures up to around 180°C.

Some components, for example, metal film and wire-wound resistors, have fairly well-defined drift characteristics, allowing the designer to make a reasonably accurate estimate of the likely change in value during the component's life. Unfortunately not all components have such a predictable performance; in some cases the drift depends upon the details of the manufacturing process, and so will vary for the same type from manufacturer to manufacturer. In other instances the drift depends upon the component value.

Where drift figures are available they can be used by the designer to relate the selection tolerance of the component to the variation which the circuit will tolerate. For example, if the designer has established that the circuit will tolerate a variation of ±20 per cent for a certain resistor value, he may specify a selection tolerance of ±5 per cent. Allowing, say, 1 per cent for the drift storage and assembly, and 1 per cent for the thermal shock of soldering on to a circuit board, the remaining variation of 13 per cent is available for drift during the working life of the equipment.

In general, information about component drift characteristics is rather sparse, and to obtain a figure for a specific set of working conditions the designer may need to interpolate or extrapolate on the available data.

A way of avoiding uncertainty is to purchase components on a basis of what is called a 'total excursion' specification. In this case the manufacturer's figure for component tolerance is not that for the initial selection but for the complete life-cycle of selection, assembly and operation for a specified period at a stated load and ambient temperature. The major obstacle to this procedure is the time taken to gather life test data, and it is much easier to apply to military or airborne equipment which has a short but arduous life, than to equipment which may last ten years or more.

5.7   De-rating to Improve Reliability

Having selected the type and value of a component, a further consideration is the so-called 'rating' of the component. This is the maximum voltage, current, or power dissipation which the component is specified by the manufacturer as capable of tolerating. Here the designer can influence reliability markedly by a suitable choice of component. The usual worst-case procedure is conventionally applied for normal commercial equipment. Thus for a capacitor we estimate the maximum likely potential which can be developed across the capacitor, then allow a moderate safety factor and ensure that the capacitor we select is rated by the manufacturer for at least this voltage. Since the cost, weight and size of a capacitor increase with the voltage rating, there is an advantage in working with only a modest safety margin.

However, capacitor life tests confirm the general validity of the fifth power law, namely that near the working voltage, the failure rate increases as the fifth power of the voltage. Consequently, for high reliability equipment it is common practice to 'de-rate' components, that is, to operate them well below the manufacturer's rated voltage. Applying the fifth power law to a capacitor working at half rated voltage gives a decrease in failure rate by a factor of 32. This considerable improvement is well worth the extra weight and cost of the component. Indeed much satellite equipment is designed using a de-rating factor of around 3.

An improvement, but to a smaller degree, can be obtained by de-rating resistors. Here the relevant quantity is power dissipation. De-rating decreases the temperature rise due to the internal heat dissipated, and thus decreases the failure rate. With low-level transistor circuits using, say, ¼ watt resistors the dissipation of most of the resistors will generally be well below ¼ watt without any particular effort from the designer, so that most of the resistors will be considerably derated. For example, a $1 \, k\Omega$ resistor requires almost $16 \, V$ to attain a dissipation of ¼ watt. Thus if we use a 10 volt supply for transistor circuits we can be certain that all resistors of $1 \, k\Omega$ and above will be derated by a factor of nearly 1·8.

Semiconductors are normally rated for maximum voltage, current and power dissipation. De-rating on all of these parameters will improve reliability, the most important ones being the junction reverse voltages and the power dissipation. Reliability can be improved by improving the cooling of the device, as well as by reducing the power dissipation.

Ideally, all data on 'limiting' or 'maximum' ratings should be accompanied by a statement of the expected failure rate, to allow the user to determine what degree of de-rating is necessary. Unfortunately, such information is not always available.

5.8   Assessed Quality Components

Having decided upon the value, type and rating of the component required, the designer has finally to specify a particular version for use in his equipment. If he has an overall reliability target for the complete equipment, he will also require some guarantee of reliability for the components. Little information is usually available for high-voltage commercial grade components, and several schemes have previously been used to provide a supply of components with assured quality. The main impetus has come in the past from the government departments which procure equipment for the services, but more recently the demand for highly reliable equipment for telephone systems, railway signalling apparatus and civil aircraft electronic systems, etc., has widened the demand for special quality components. It is obviously uneconomic for all of these purchasers to write their own specifications for components, each requiring different kinds of life testing, etc., so starting in 1961 an attempt was begun in the UK to establish a common source of components of assessed quality, suitable for both military and civilian applications.

The first step was the establishment of committee chaired by Rear-Admiral G. F. Burghard which produced the so-called 'Burghard scheme' for a system of common standards for electronic parts for military and civilian use. The committee's final report in 1965 was accepted by government and industry, and the British Standards Institution assumed responsibility for preparing the necessary specifications.

The basic document which describes the system and specifies the way in which parts are approved is BS 9000. BS 9001 gives general rules and tables for production sampling, and BS 9002 gives brief details of all components which have been approved and their manufacturers. These documents apply to all components within the scheme.

Owing to the wide variety of electronic components, it is necessary to produce separate specifications for each family of components, and in certain cases for particular sub-families. These are called generic specifications and include families such as

BS 9010 – Transmitter tubes
BS 9050 – Cathode-ray tubes
BS 9070 – Fixed capacitors
BS 9110 – Fixed resistors
BS 9300 – Semiconductor devices
BS 9400 – Integrated electronic circuits
etc.

Since the same test procedure cannot be used for all semiconductor devices, detailed specifications have been issued for particular sub-families, for example

BS 9301 – General purpose signal diodes
BS 9302 – Switching diodes
BS 9304 – Voltage reference diodes
BS 9305 – Voltage regulator diodes

etc.

In order to provide some independent supervision of the manufacturer's testing and quality assurance procedure, the Electrical Quality Assurance Directorate of the Ministry of Defence are responsible for approving manufacturers, stockists and test houses, and for ensuring that the appropriate testing and quality control is maintained. By the middle of 1973, approval had been granted for fifty-five manufacturers, twenty distributors and eleven test laboratories, twenty-three generic specifications and forty-five detailed specifications had been issued. Also just over 150 semiconductor specifications from the former CV 7000 series have been adopted under BS 9300.

The tests required cover physical and mechanical properties as well as electrical characteristics; for example, a voltage regulator diode to BS 9305 is tested for

> solderability of leads
> failure (bending) test of leads
> temperature cycling: 55°C to 150°C and damp heat
> vibration and acceleration: 150–2 000 Hz at 20 g
> > followed by acceleration test at 20 000 g
> electrical endurance: at least 160 hours at maximum dissipation

In addition to these short-term tests which are conducted at regular intervals, long-term tests which may continue for up to 8 000 hours are also conducted. A unique feature of the BS 9000 scheme is that the results of the long-term and short-term tests are issued in the form of Certified Test Records (CTR) which the customer can examine to assess the reliability of the component. These have to be issued every six months for components supplied to detailed specifications in the general and special application categories. Thus when components have been supplied by a manufacturer to a BS 9000 specification for several years, the CTRs afford not only information about component reliability but also the consistency of the production process. Furthermore, as the CTRs accumulate the total number of component hours of testing increases, and so the statistical accuracy of any failure rate calculation based upon them also increases.

The advantages of a single and consistent reliability assurance system such as BS 9000 are now well appreciated, and several Continental countries have shown an interest in the scheme. A European committee called CENELEC comprising fourteen EEC and EFTA countries was set up to harmonise and standardise specifications for electronic components in 1972, replacing an earlier committee with a somewhat smaller membership which had operated for some years. The members of CENELEC are the national electrotechnical committees of the participating countries. A similar committee called CENEL deals with electrical standards.

Since the BS 9000 proposals were the first to be formulated in any detail, the European standards were based on these. The key document is CENEL 1, which is similar to BS 9000. Future BS 9000 activities will involve co-operation with CENELEC, and the latest version of BS 9000 has been modified slightly to bring it into line with CENEL 1.

At present only a few European generic specifications have been issued, but ultimately it is expected that they will cover the complete range for which the BS 9000 system is intended, and the two sets of standards will be harmonised. The long-term aim is that components purchased from any European source which is approved under the CENEL scheme will have been tested to the same specifications and quality standards. At present only limited attempts have been made to spread the scheme outside Europe, but Canada now has a system similar to BS 9000, and so has Japan. An IEC committee is also considering a truly international scheme but this is likely to be some years from fruition.

5.9 Mechanical Design

The final aspect of designing for reliability is the provision of suitable arrangements for mounting the component and if necessary controlling its environment. The procedure required depends very much upon the surroundings in which the equipment must operate. Fixed, ground-based equipment such as computers and electronic telephone exchanges normally have favourable environments, and components in them do not require any particular protection or any complicated cooling system apart from a moderate amount of forced air circulation. By contrast, portable equipment, especially that made for worldwide use, may suffer from dust, salt-laden moist air, severe variations of temperature and considerable vibration and shock. A substantial degree of protection can be obtained by encapsulating modules of the equipment in resin, but such units then become plug-in non-repairable items and the cost of a complete set of spares may become considerable. Also encapsulation generally degrades cooling, so it may be necessary to embed heat sinks or heat conductors in the encapsulating medium to improve cooling.

In many environments, particularly aircraft and military vehicles, it is desirable to use anti-vibration mounts to isolate the equipment from its surroundings. Similarly, some thermal barriers may be required if equipment must be mounted near to surfaces at a high temperature.

The designer has two lines of defence against a hostile environment. Firstly the component and the module in which it is mounted can be constructed and treated so as to resist the effects of dust, corrosive atmosphere, vibration, etc., and secondly the complete assembly can be enclosed in a housing which also resists to some extent the environment.

A particular difficulty which arises in airborne equipment is the reduced air pressure at high altitudes. Electronic equipment is usually mounted outside the pressurised area and thus must operate in the ambient pressure. This causes two main problems, poorer cooling and reduced flash-over voltage. Extra cooling arrangements may be needed to ensure that components do not overheat, and extra insulation, or increased conductor separation may be required in all circuits carrying, say, 200 V or more. The usual area which requires most attention is the power unit for cathode-ray tube supplies. Another possible source of trouble may be the rf circuits of radio transmitters. In all airborne equipment a proper low-pressure environmental test is essential to validate the design.

5.10   Protection against Interference and Noise

Just as equipment in a hostile environment needs to be protected to some degree against the mechanical disturbances introduced by its surroundings, so it may also need protecting against electrical disturbances. These can arise from electrical or magnetic fields caused by neighbouring machinery or cables, or from e.m.fs induced into external leads. Interference from external fields can usually be reduced to tolerable proportions by screening, the metal case normally used for mechanical protection being often sufficient. However, this does not screen against fields generated internally by the equipment itself, and protection in this case is usually a matter of careful layout together with perhaps screening of certain critical parts of the circuit.

Modern transistors and integrated circuits, being both of wider bandwidth and using smaller signals than the valves they replaced have seriously accentuated this problem. Typical integrated logic circuits may have a bandwidth from d.c. to, say, 70 MHz or more, and require signals of less than a volt to make them change state. Thus it is important to avoid much electrostatic or magnetic coupling between adjacent conductors. 'Earth-plane' techniques in which all signals leads are close to, or sandwiched between, conducting planes held at earth potential are helpful for this. They are generally used in the form of multilayer printed circuits in modern digital computers. When using high-speed digital elements, all connections longer than a few inches need to be treated as transmission lines and correctly terminated to avoid spurious operation due to reflected signals.

Connections to peripheral units, up to, say, 50 ft long, are operated differentially, using special driving and receiving amplifiers. The drivers operate at a higher current level than conventional logic, and the receiving amplifiers have differential input stages with high common-mode rejection for signals of, say, ±3 V.

It is normally possible to avoid interference troubles in self-contained computer installations since all of the equipment is under the control of one contractor, usually the manufacturer of the equipment. Much greater difficulties arise with on-line digital equipment used to control machine tools and industrial processes. Here sensing signals from the plant or process may have to travel several hundred yards, perhaps traversing areas where heavy electrical plant is working, and considerable induced interference voltages may be injected into the cables. For this kind of situation special high-level integrated circuits have been developed, which require 12 or 15 V rather than 5 V supplies, and have a much higher noise immunity than normal TTL circuits. Complementary MOS logic is also useful for these applications, since it can operate with supply voltages of around 20 V, and has a noise immunity of about 40 per cent of the supply potential.

Some short interference spikes can be removed by low-pass filtering, but the amount of filtering which can be used depends upon the pulse rate of the incoming signals. The cut-off frequency of the filter must be chosen so that the edges of the pulses are not excessively degraded, a rough rule being that the product of bandwidth and rise-time should be about 0·35. Thus if data is arriving with a 10 kHz pulse rate, a complete cycle takes $100\mu s$. Allowing, say, $30\mu s$ for the top and bottom of a pulse of 1:1 mark-to-space ratio leaves $20\mu s$ for each edge. The corresponding bandwidth is

$$\frac{0.35}{20 \times 10^{-6}} \text{ Hz } = 17.5 \text{ kHz}$$

A major difficulty in these long circuits is the common-mode interference which may be too large for normal differential receiving amplifiers to tolerate. A method of overcoming this is to use optical isolators. These use a light-emitting diode and a photo-diode in close proximity, the only signal coupling being via the light beam. Thus the transmitter may be operated with no electrical connection of any kind to the receiver, and common-mode interference is of no consequence.

Where the signals being handled are relatively slow, with pulse rates of, say, 50 Hz or less, high-speed electromechanical relays can be used to isolate the computer circuits from the external leads.

Interference can be injected into electronic equipment through any line leading to external plant, including, of course, the supply lines. Switching of adjacent loads can cause severe transients on power supply lines, and the growing use of thyristor controllers and regulators has increased the sources of short high-voltage transients which are particularly liable to cause faulty operation in digital equipment.

Measures which can be taken to diminish mains-borne interference include the insertion of filters in the mains leads, the use of an earthed electrostatic screen round the primary of mains transformers, and, if accessible, energy absorbing networks across all switches and contactors which control other loads on the mains. The transformer leakage reactance helps the filtering on all transformer-coupled equipment, and a more difficult problem arises with directly-fed apparatus.

This became apparent with early transistorised equipment for aircraft. Since all civil airliners have a 28 V d.c. supply, this was considered the most convenient source for the equipment. Unfortunately the transistor failure rate was much above expectations, and it was discovered that when switching heavy inductive loads on the supply, short pulses of over 500 V were produced. Such pulses, lasting only a few microseconds, can destroy transistors rated at 30–40 V, and the filter on the supply must attenuate them by over two orders of magnitude to ensure satisfactory operation. Sometimes a high-current Zener diode is connected across the supply to the transistor circuits, having a breakdown voltage, say, 20 percent above the normal supply potential. This will bypass any high-voltage surges and so ease considerably the task of the supply filter.

Similar problems beset linear amplifier circuits; these can usually be arranged to operate correctly and stably without difficulty with short output and input connections, but may suffer from spurious signals and interference when the connections are extended. One difficulty is caused by the extremely wide bandwidth of modern planar transistors. Even though the designer may intend to produce a low-frequency system, parts of the circuit may easily amplify or oscillate at frequencies of tens of MHz or above. Thus when the input leads are extended, unless carefully screened these may pick up a wide range of interference which may be amplified or rectified by subsequent transistor circuits. The same may happen even if only the output leads are extended, since it is customary to use overall feedback and there is thus a direct, although attenuated path from the output to the input of the amplifier. Consequently any signals induced in the output leads may be passed to the input of the amplifier. If they are large enough to cause overload and thus intermodulation with the wanted signal, no subsequent filtering can eliminate them without affecting the wanted signal. This kind of interference may arise only at a particular frequency, and with a certain arrangement of input leads and earthing arrangements, and it is thus quite difficult for the designer to be sure that no such combination can cause trouble. In general it is good practice to ensure that amplifier bandwidth is no greater than that needed to handle the normal signal, and that as far as possible multiple earths are avoided particularly for input circuits.

Some complicated control systems pose severe problems in this respect, since it may be necessary to combine, while retaining the d.c. component of most signals, various feedback signals derived from different parts of the system with the demanded input and the system output. It may not be possible to arrange a common reference potential for all of these circuits, and so they may be used to modulate a carrier wave. The resulting modulated a.c. signals can be combined in a multi-winding transformer without direct connections between the various signals. An alternative solution suitable for frequencies up to, say, a few kHz is to use a magnetic amplifier as a mixing circuit. Each input is connected to a separate control winding, and as before no direct connection need be made between them.

5.11 Using Redundancy

Although a basic single system is the designer's normal choice for economy in space, weight, power consumption, etc., a demand for very high system reliability may require some degree of redundancy. The improvements in reliability theoretically attainable using various arrangements have been discussed in chapter 3. The designer's task is to select the most suitable arrangement and to make sure that the theoretical improvement is in fact obtained.

One of the major difficulties is to ensure that the redundant channels are as far as possible independent, so that a fault on one of them does not alter the probability of a fault on the others. It may not be possible to ensure this; if not, the common factor must be analysed and the probability of this causing a failure must be added to the failure probability of the redundant part of the system. Any single error- or fault-detecting circuit clearly comes into this category. Also, although the system may use majority voting with replicated voting elements, the last voting element must be a single one, since only a single output is required. Consequently the overall system reliability can never be greater than the reliability of this final voting circuit, and it is necessary to make this final circuit as reliable as possible. With digital systems producing an electrical output the logical function of the circuit requires four gates, to generate from the three inputs, say, A, B, C, the expression

$$X = A.B + B.C + C.A$$

The only way to enhance the reliability of the circuit is to use redundancy at a lower level, that is, at gate or component level, since the expression for X cannot be simplified.

With analogue circuits the final switching or combining unit may be somewhat more complicated, particularly if the output is an electrical signal. If the output is a mechanical movement or rotation, it may be possible to retain the multichannel configuration until nearly the system outputs, and employ a very simple arrangement for the final voting or combining circuit.

A particular example of this occurs in some aircraft blind-landing equipment using triplicate redundancy. Here the 'voting' or combining element is a single shaft to which are coupled the three driving motors which are the final components in the three channels, and the control surface to be moved. The nett actuating torque is the sum of the torques of the three motors. The three motor driving amplifiers have current- and thus torque-limiting facilities to protect the motors. Given the maximum torque which the motors can develop, the designer can easily proportion the shaft to have a torque capacity much in excess of this.

The probability of failure of the final combining element is then the extremely small probability of failure of the shaft. This is usually less than the probability of a design fault, that is, an error on the designer's part which causes him to select an unsuitable component. This is a human rather than an equipment failure which should be discovered during prototype tests. Its likelihood can be much reduced, like component faults, by redundancy, that is, by ensuring that all design calculations are checked by engineers other than the original designer.

The possibility of a design fault escaping prototype tests and being present in the production models of equipment is one we have hitherto neglected, but as components become more and more reliable it acquires greater relative importance. It is generally disregarded in most reliability predictions, but where these involve extremely small probabilities, such as in the highly redundant control circuits for nuclear reactors, one safety assessment authority has suggested that no unit failure probability less than 10^{-6} should be accepted, since whatever reliability the equipment might possess, the figure of 10^{-6} should be added as some recognition of the probability of a human designer's error, or failure to visualise some particular failure mode.

An obvious common factor in all redundant systems is the power supply. This must either be extremely reliable (more so than the complete system is required to be) if it is common to all channels, or must be replicated like the rest of the equipment. In order to obtain maximum reliability such supplies should if possible be derived from physically separated sources, so that, for example, a fire in one machine cannot cause a fault in a second source of power.

5.12 Maintained and Non-maintained Systems

The design of a redundant system depends to a considerable degree on the user's ability to maintain it, or perform some corrective action. If the system is inaccessible and no human maintenance can be performed, the main consideration in the design and assembly is the initial checking and testing procedure. Once the user has checked that the system is working correctly at the start of its operating period nothing further can be done.

If, however, the equipment is used in an accessible situation, regular checks and repairs are possible, and the user is often concerned more with the availability than any other reliability parameter. This can be increased without using components of better inherent reliability by reducing the repair time. This means having readily available (and perhaps built-in) diagnostic test equipment, monitoring points and system status indicators. All of these enable the repair technician to locate the fault rapidly. The second requirement is adequate provision of spares and rapid means of replacing a faulty sub-assembly.

In any scheme of redundancy some systematic way of testing for, or automatically indicating, faulty elements is most important if full benefit is to be obtained from the redundancy; without this facility the redundant system may ultimately become worse than a single channel system. For in the course of time the redundant system will accumulate faults which are masked by the redundancy until a multiple fault occurs at some point, so causing a system failure. This fault is repaired and the system returned to service. Ultimately nearly all of the redundancy will be ineffective because of faults which have not yet caused system failure, and nearly every subsequent fault will stop the system. If the system is, for example, a triplicate arrangement, there will be rather more than three times as much equipment as in a single channel system, and so, for the same component failure rates the ultimate system failure rate will be at least three times greater for the triplicate system.

Of course, a considerable time has to elapse before this situation develops, but it must eventually arrive in all redundant systems unless every fault is cleared when the system goes down. One consequence of this is that the full benefit of redundancy is obtained in a non-repairable system only during the initial part of its life, for a period considerably less than its MTBF. The calculation of section 3.18 shows that in fact the MTBF of a triplicate system with no maintenance and an ideal voting circuit is only 5/6 of the MTBF of a single channel system.

5.13 The Use of Fault Indicators

In any automatic scheme for switching in stand-by equipment a fault indicating unit must be provided to initiate the switching, which can also warn maintenance staff that one part of the system has failed. With majority voting arrangements, however, some extra apparatus is needed to detect a disparity between two channels. This is no more complicated than that required for majority voting, and can enhance the system reliability considerably by warning the maintenance staff of all single faults which have not yet caused a system failure.

For a maintained system the important quantity is usually availability rather than reliability. Thus for a single system with a mean time between failures of M, and a mean repair time R, the availability, that is, the average fraction of the switch-on time that the system is working, is given by

$$A_1 = \frac{M}{M+R} = 1 - \frac{R}{R+M}$$

This is the steady-state value attained after the initial transient dies away. The unavailability, or the proportion of time the system is inoperable, is

$$U_1 = \frac{R}{R+M}$$

Just as redundancy decreases the probability of a system fault, it also decreases the unavailability, since, for example, with a twin parallel system a failure occurs only when one channel develops a fault while the other channel has failed and is being repaired. The second fault must occur within a time $\pm R$ from the first.

The steady-state unavailability of the twin system is

$$U_2 = \frac{2R^2}{(R+M)^2}$$

approximately, if R^2 is considerably less than M^2 as is generally the case.

The assumption here is that no change in repair time occurs when there are two channels to maintain.

This unavailability is a constant times the square of the unavailability of a single system, so full benefit is being obtained from the redundancy.

The result for a triplicate scheme with majority voting is similar, since two channels must fail before the system becomes inoperable, apart from the numerical constant.

The above expressions for availability are not fully realised in practical systems for several reasons. Firstly the repair time is assumed, like the time between failures, to have an exponential distribution. This is not in accord with most experimental evidence which suggests that there is a minimum time required for running through diagnostic and test routines before any repair work can start. Also no fault indicator is perfect; the possibility of its developing a fault must be recognised in the detailed analysis of the system. The performance of the system will also depend upon the amount of effort available for repair.

The final result of introducing these additional factors is that the system is not easily susceptible to analysis and the quickest estimate of availability may be a computer simulation of, say, 1000 faults.

5.14   Digital Fault Indicators

The simplest majority voting arrangement is obtained with a triplicate
system, and involves only four gates. A basically similar arrangement
will also act as an error detector. This is required to detect a disparity
between one channel and the other two. Thus the combinations of
input signals to which it must respond are 0, 0, 1 and 1, 1, 0 but in
any order.

If the three inputs to the error detecting circuit are A, B, C and the
output is X, there are six possible combinations which must produce an
output X, given by

$$X = \bar{A}.\bar{B}.C + \bar{A}.B.\bar{C} + A.\bar{B}.\bar{C}$$
$$+ A.B.\bar{C} + A.\bar{B}.C + \bar{A}.B.C$$

This expression can be simplified to

$$X = A.\bar{B} + B.\bar{C} + C.\bar{A}$$

or

$$X = \bar{A}.B + \bar{B}.C + \bar{C}.A$$

A simpler expression can be derived by considering the remaining two
combinations which represent correct operation. These are 0, 0, 0 and
1, 1, 1. Thus correct operation is denoted by

$$Y = \bar{A}.\bar{B}.\bar{C} + A.B.C$$

and incorrect operation by the complement of this, namely

$$Z = \bar{Y} = \overline{(\bar{A}.\bar{B}.\bar{C} + A.B.C)}$$

This can be implemented by two AND gates, an OR gate and an
inverter, or by three NAND gates and an inverter if both the inputs
$A \times B \times C$ and their complements are available. If not, three extra
inverters are required.

In all of these error detecting circuits a '1' output denotes an error
and a '0' output denotes a correct operation.

In addition to the hardware error check described above, it is often
possible to inject known signals into the input stream of a digital system
and check that they are handled correctly. This, of course, involves
periodic rather than continuous checking, but does not require any
additional hardware error detectors. The introduction of this kind of
check detracts to some degree from the data handling capability of the
system since the time spent on checking is not available for data

handling. However, it may be possible to fit the checking in between
bursts of data handling in some situations. For example, computers
used for process control and data collection normally have a set of
programs of differing priority which control their various activities.
At the bottom of the priority list can be the test program which
exercises and checks all of the facilities of the processor. This test
program will be called in only when no other action is required from
the system.

5.15   Analogue Redundancy

One consequence of the reduction in cost of digital logic has been an
acceleration of the trend towards digital rather than analogue signal
processing. Despite this there are still many situations in which information must be handled in an analogue form, and where reliability is
important some form of redundancy is desirable.

One possible method is to have some switching mechanism whereby
a faulty channel is switched out of operation and replaced by a good
one. This requires some arrangement for continuously sensing whether
a channel is operating correctly. Since the information coming along the
channel has a considerable element of randomness in amplitude and
frequency, it is necessary either to use some predictable feature of the
signal or to add to the channel some known and constant factor. This
feature can be sensed at the receiver and used to indicate whether the
channel is working or not.

As an example, the carrier wave in a conventional AM transmission
is not affected by the magnitude or frequency of the modulating signal.
Its amplitude can be measured at the receiver and used to indicate that
the channel is working. A suitable threshold value has to be used to
ensure that in the absence of the carrier the noise and interference
picked up by the receiver is not confused with the wanted signal. The
threshold is set well above the noise level, and any signal above the
threshold for more than a minimum time is accepted as a valid signal.

In an FM system the total signal amplitude should also be constant,
and this can be used to check that the received signal is present. In
carrier telephone systems on the other hand single sideband suppressed
carrier modulation is used. In order to have some constant factor
independent of the modulation it is then necessary to add a 'pilot'
signal to the transmitted signal. This is usually located just outside the
signal frequency band so that it is handled by all intermediate amplifiers

along with the normal signals, and can be filtered out and rectified at the receiver. In addition to indicating that the channel is working, the pilot tone can also be used for gain regulation. Slow changes in pilot tone level are corrected by the regulator, but an abrupt drop in level causes the system to switch to an alternative channel.

There are some situations in which the introduction of a pilot signal would be inconvenient, and for simple units such as amplifiers a balancing process employing the normal signal has been used. Here the output signal is passed through an attenuator which has a loss equal to the amplifier gain. The output of the attenuator should thus be equal in amplitude to the input signal. The difference between the two is amplified and used as an error indicating signal. If the amplifier gain increases or decreases, the balance between the two signals is disturbed and an error is signalled. The same happens if the amplifier overload point drops so that large signals are distorted. The comparison between input and output is somewhat more complex than described above, since, for example, a time delay may be needed in the input signal to balance the time delay of the signal through the amplifier. Also, of course, the error will not be detected unless a suitable input signal is present.

5.16   Parallel Redundancy

For situations where no break in communication can be tolerated, or at low level where complicated switching circuits would be unsuitable, parallel redundancy can be used in which all channels are operating continuously, and their outputs are combined to give a single output. The combining circuit must be able to disconnect a faulty channel, so that, for example, it cannot short-circuit the other channels, and each channel should be capable of supplying the output alone. These are exacting requirements and usually one must accept some small change in performance as channels develop faults. One solution is to use a multi-winding transformer as the combining unit, with some series resistance in each channel output to prevent a short-circuit fault. If feedback is taken from the transformer the change in output level due to a short-circuit channel output can be minimised. However, this arrangement is suitable only for a.c. coupled systems, whereas most analogue systems require the d.c. signal component to be retained.

A different procedure can be used for directly-coupled amplifiers. Almost all faults in these will shift the d.c. level of the output significantly, usually to one or other of the limiting potentials. Thus if such

amplifiers are used in IF strips or similar positions where they are
amplifying a.c. signals only, the d.c. component at the input can be
removed by capacitance coupling, and any substantial change in the
d.c. level of the output will reveal a fault in the amplifier. With single-
ended output stages which are operating effectively in parallel, a high
potential output caused by an open-circuit or cut-off output transistor
will not diminish the gain greatly. A short-circuit or saturated transistor
will on the other hand effectively short-circuit the common output
of the redundant amplifiers unless disconnected. The circuit shown in
figure 5.2 can be used to disconnect automatically any amplifier with

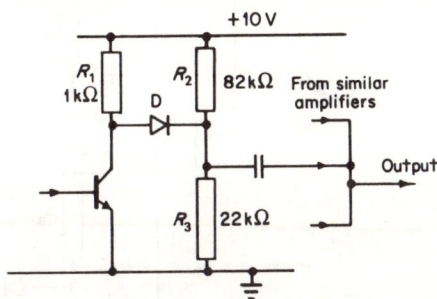

Fig. 5.2 Amplifier output circuit

an exceptionally low output level and leave the remaining amplifiers
to supply the load. The arrangement shown will disconnect the transistor by reverse biasing the diode D when the transistor collector
potential falls below about 2·6 V. The quiescent collector voltage should
be about 6·3 V, assuming a symmetrical input signal. This arrangement
is most effective when the amplifier is used with voltage feedback which
reduces the output impedance. If we have a duplicate system without
feedback, any failure which causes an open-circuit or a cut-off output
transistor will still leave the resistor R_1, shunting the other amplifier and
thus reduce the output voltage by nearly half with the load resistor of
1 kΩ shown in figure 5.2. On the other hand, if voltage feedback is used,
the output impedance of the good amplifier may be only, say, 100 Ω.
The extra 1 kΩ shunting this from the faulty amplifier will then affect
the output voltage much less.

The arrangement shown in figure 5.2 is more convenient for duplicate
or triplicate schemes; if more amplifiers are used in parallel the change

in output, if they all fail but one, may be excessive. An alternative scheme for triple redundancy is shown in figure 5.3. Here the output always follows the middle voltage if all three differ, or takes a majority vote if two of the inputs are the same.

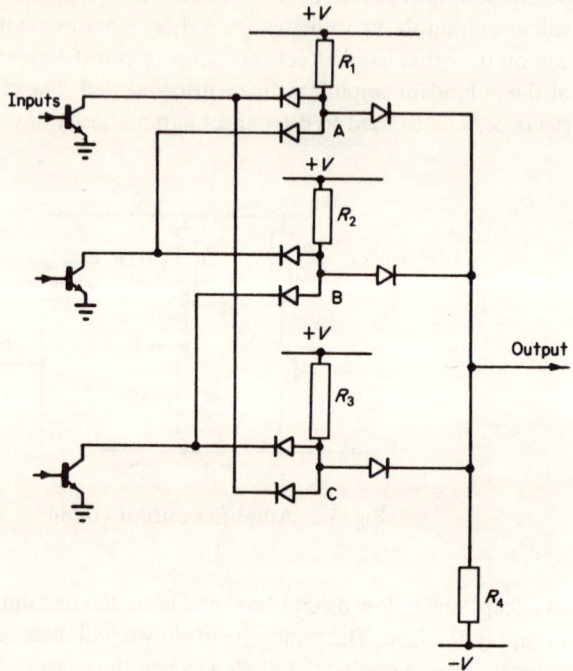

Fig. 5.3 Analogue majority voting circuit

If the inputs are all at different potentials, the first set of gates compares the inputs two at a time, and each gives an output corresponding to the smaller of the two input voltages. Thus two of the junction points A, B, C, will have the potential of the least positive input, and the third that of the next most positive input, that is, the input which is intermediate in potential between the other two. The second gate selects the largest of the three signals at A, B, C, that is, the intermediate of the three input signals. A similar argument when two inputs are at the same potential will conclude that the output is a majority vote on the inputs.

This circuit was first proposed for the voting on a triplicate autopilot system for aircraft. In addition to taking a majority vote on the inputs, it will pick the correct channel when two inputs are faulty, so long as one input is hard-over to earth and the other hard-over to the positive supply. If, however, two inputs are hard-over in the same direction, the circuit cannot select the remaining correct signal.

There will be a small change in d.c. level between input and output, since, although the diodes may be matched, it is necessary to use a higher current level in R_1, R_2, R_3 than in R_4. The arrangements of figures 5.2 and 5.3 are particularly useful for integrated circuit amplifiers since these are almost without exception directly coupled.

5.17 'Fail-safe' Circuits

In all previous discussion and calculation we have assumed that in digital circuits a failure to a '0' output and a failure to a '1' output are equally damaging, and no distinction has been made between them. This is appropriate for all computing hardware and equipment handling digital data which comprises a mixture of '0' and '1' digits. It is generally not true, however, for monitoring or safety or alarm circuits which spend nearly the whole of their lives emitting one signal, and change to the other only in faulty or unsafe conditions. The penalties for failure to '1' and for failure to '0' may be quite different, and it is necessary to estimate separately the probabilities of the two events. This form of analysis has been developed mainly in connection with nuclear reactor safety systems, but is equally applicable to other systems where safety is paramount, such as railway signalling.

If we consider that a monitoring unit which is checking the value of some physical quantity emits a '0' for safety, and a '1' for danger, we can recognise two failure modes.

The first is the so-called 'fail-safe' error, in which the output changes to '1' when the physical quantity is still in the safety zone. The second is the 'fail-danger' error in which the output stays at 0 when the measured quantity is in the danger zone. In a safety system the first error produces a false alarm, and the second an alarm failure.

In many cases redundancy is used to improve the reliability of the safety system. Again more detailed analysis is required than we have so far used, since some kinds of redundancy will improve the fail-safe performance, and others will improve the fail-danger performance.

For example, using two monitoring units an AND combination of the two outputs requires both units to emit a '1' to indicate a fault. This will decrease the chance of a false alarm, but increase the chance of an alarm failure. An OR combination of the outputs will do the converse. If, as usual, we require greater reliability for both conditions, majority voting is appropriate, since with a triplicate system, for example, any action requires agreement between two of the units.

In a nuclear reactor used for generating electricity, a fail-safe fault will shut down the reactor unnecessarily, so reducing the station's earning capacity until the situation can be restored to normal. However, if a reactor fault occurs when the safety system is in a fail-danger state, the expected shut-down of the reactor may not occur until the operator recognises the fault and intervenes manually. The consequence may be reactor damage and the escape of radioactive material, a much more serious and costly event than the fail-safe error.

In the design of all instrumentation and the safety circuits it is thus necessary to estimate the probabilities of fail-safe and fail-danger separately. This requires a more detailed set of fault statistics than are often available, since at component level it requires separate recording of short-circuit and open-circuit faults. This information is usually recorded during life tests, but much published information lumps together all types of failure in a single total.

In order to produce an acceptably reliable reactor safety system, triplication of monitoring units with majority voting is generally used, operating at two levels in critical areas. This involves nine monitoring units grouped in threes feeding three majority voting circuits whose outputs feed a further voting circuit.

In addition to this redundancy the monitoring units are carefully designed so that as far as possible component failures will result in a fail-safe rather than a fail-danger condition. This is an extra constraint on the circuit designer and leads to unusual circuit configurations. For example, an important factor is fuel element temperature, measured by thermocouple and amplifer. Failure in a d.c. amplifier normally causes the output stage to limit or saturate in one direction or the other. In order to produce the required fail-safe characteristics, high gain d.c. amplifiers are avoided in favour of chopper amplifiers. Nearly all faults will then produce zero output which can be arranged to indicate a fail-safe condition.

The final improvement in reliability can be obtained by using better logic elements in the safety circuits themselves. For the low-level circuits coupled to the monitoring unit outputs conventional electromechanical relays were originally used. These are now replaced by devices similar to magnetic amplifiers, but using multi-aperture ferrite cores ('laddics'). These have no moving parts, and a single laddic can provide a majority decision vote. The safety circuits thus comprise fewer, more reliable components and are consequently significantly more reliable than the earlier relay circuits.

Although the above developments have mainly been stimulated by the demands of the nuclear power industry, the same 'fail-safe' principles and design techniques are relevant in many other areas, particularly where human safety may be involved.

The above discussion of 'fail-safe' techniques is based upon an assumption that the output of any monitoring unit can have one of only two possible values, as in conventional binary logic circuits. Thus a failure must produce an output corresponding to one or other of these values. If, however, it were possible to construct three-state logic elements, it would be possible to assign the normal binary values of '0' and '1' to two of the states, leaving the third state as the failure mode. These could be, for example, $+V$ for '1', $-V$ for '0' and earth for failure. The aim of the circuit designer in implementing fail-safe operation would then be to make as many of the possible failures modes as he could produce a zero output potential. This state could be detected by a relatively simple fault indicator.

As yet a good deal of theoretical study of tri-state logic has been achieved, together with some implementation, but the gates are somewhat complex and require close-tolerance components, and are not much used.

Some medium scale integrated circuits have been produced which have in fact three possible output states, namely '1', '0' and 'turned off'. In the first two states, the output impedance is low and the output voltage is dictated by the circuit. In the third state a control signal is changed, the output circuit is effectively turned off, so that the output impedance is very high, and the output voltage is dictated by some other circuit connected in parallel.

This is not true tri-state logic since there are only two possible signals for each logic input, the OFF state being controlled by a different GATE or ENABLE input. However, it may provide a starting point for the future development of tri-state logic which has a distinct field of application in the design of fail-safe switching systems.

5.18   Integrated Circuit Redundancy

The major development in the construction of electronic equipment during the 1970s has been the growing use of integrated circuits. From the first devices which incorporated a dozen or two transistors in simple logic circuits, these have developed into much larger units embodying many thousands of transistors and providing both analogue and digital functions. Currently over 60 000 storage cells can be fabricated on a single digital chip, and analogue circuits such as a phase-locked loop, or the entire active circuits for an A.M. receiver can be mounted in a single package.

The resulting circuits are generally more reliable than their predecessors, when used in a non-redundant mode. However, the low-level redundancy which is most effective with discrete component circuits can no longer be employed for applications needing high reliability, since the basic unit which can be replicated is an integrated circuit which incorporates a much higher level of complexity.

This argument is based upon a simplified calculation of failure rate which assumes that the basic system comprises a number of components large enough to enable voting circuits to be introduced at many points along the system. Further assumptions are

- that all voting circuits have the same failure probability P_V
- that a voting circuit can be introduced at any point in the basic system without requiring any additional components.

We will apply these assumptions to a triple redundancy scheme using three equal systems each having a failure probability of P_0.

If each system is divided into N equal sections, each followed by a voting circuit, the failure probability of the combination is

$$P_1 = \frac{P_0}{N} + P_V$$

For the triplicated combination, the failure probability is

$$P_r = 3P_1{}^2 = 3\left(\frac{P_0}{N} + P_V\right)^2$$

neglecting terms involving $P_1{}^3$. There are N of these combinations in the complete redundant system, so that the failure probability of the redundant system is

$$P_s = NP_r = 3 \left(\frac{P_0}{N} + P_V \right)^2 N$$

$$= \frac{3P_0^2}{N} + 6P_0P_V + 3P_V^2 N$$

The variable in this expression is N, the number of sections, and the value of N giving a minimum probability of failure is found by equating dP_s/dN to zero. Thus

$$\frac{-3P_0^2}{N^2} + 3P_V^2 = 0$$

whence

$$P_V = \frac{P_0}{N}$$

This result can be applied equally to all-digital or all-analogue systems provided that all voting circuits used have similar failure probabilities.

The result shows that to produce the most reliable replicated system, we must divide the basic system into sections which have the same failure probability as that of the voting circuit. Since the voting circuits will generally use components and technology similar to those embodied in the system they are attached to, this means that the complexity and component count of the system section is similar to that of the voting circuit.

Since a majority voting circuit for a triplicated logic system can be assembled from three two-input AND gates and a three-input OR gate, for optimum redundancy a logic system will need very many voting circuits. Assuming that most gates have two or three inputs, about $N/4$ voting circuits would be needed in an N-gate system.

The advent of complex integrated circuits such as microprocessors with the equivalent of several thousand gates on a single chip, and only 40 or fewer pins connecting them to external circuits means that optimum redundancy cannot be implemented. In this case we must make do with fewer than 40 voting circuits (1 for each signal pin) whereas the optimum system might require 500 to 1000.

However, the minimum failure probability computed above does not depend very critically on N, so that considerable benefit can be obtained with fewer than the optimum number of voting circuits.

A further difficulty in replicating microprocessors in the problem of synchronisation. Usually each system will have its own clock supply derived from a crystal-controlled oscillator, and these cannot be held in synchronism for long. Thus if the scheme of replication calls for the three or more processors to vote on the outcome of each instruction, they must share a common clock supply to retain synchronism. We then have the difficult task of voting on the outputs of the separate oscillators which are all running at slightly different frequencies, perhaps selecting that which runs at a frequency nearest to the mean.

Alternatively, the processor may be allowed to execute a few instructions asynchronously and then wait. As soon as all processors have finished a vote is taken on their current outputs and this value is used subsequently by all. Any processor consistently producing false results can be disconnected and the voting procedure altered.

The scheme used depends upon the environment, and particularly whether maintenance is possible. If so, fault indicators are required and the moment a faulty processor is detected it should be replaced by a new one. This usually means plugging in a new board containing processor, stores for program and data, and some input/output packages. Additional program segments are required to allow the new processor to synchronise itself with the others already working, and load sufficient data into its data store to compute correctly.

If on the other hand the processors are installed in an inaccessible section of a train, aircraft or plant, no maintenance can be performed in service and the equipment must disconnect sections which indicate permanent faults and reconfigure itself to give maximum reliability.

For example a control system containing three processors, and producing relatively few outputs can employ majority voting circuits to supply each of the outputs. As soon as any processor permanently disagrees with the other two it can be disconnected. The voting circuit is then switched so that one of the remaining processors supplies the output alone. The second monitors the action of the first; any divergence indicates a fault and both processors are then disconnected. In most cases the system would then revert to manual control.

In addition to physical replication as described above, temporal redundancy can be used where a signal is derived from a noisy channel or transducer. If time permits a dozen or so readings of a particular value can be taken, the average evaluated, and any values too far from the mean discarded. The mean of the remaining readings can be used to give a better estimation of the value than can be found from a single sample.

The introduction of semiconductor storage, which appears more liable to occasional random errors than the preceding ferrite core store has stimulated interest in error-checking and in some cases error-correcting coding for store modules.

Although nearly all current integrated circuits are non-redundant, in some devices where a particular circuit is used many times on the same chip a degree of redundancy has been incorporated to increase the yield of usable packages. The main application is to digital stores where several spare colums of cells are fabricated on large storage chips. If only one or two columns are faulty these can be isolated and the spare columns connected in their place by modifying the connections on the chip. This process allows many otherwise useless chips to be rendered usable.

The preceding discussion has been largely confined to digital circuits, but similar problems occur in replicating analogue circuits. These are generally less intractable since analogue integrated circuits are usually much smaller than digital integrated circuits, and the voting circuits more complex.

Also for operational reasons it is often preferable to replicate complete units rather than to subdivide them. Thus in the highly redundant safety systems for nuclear power reactors, complete instrument channels such as neutron flux detectors, period detectors, temperature measuring units, etc., are replicated. Each individual unit is non-redundant but is as far as possible a fail-safe configuration.

5.19   Future Trends

The history of reliability studies over the last 40 years has included the replacement of less reliable components by more reliable ones, and the development of new devices which have much greater inherent reliability. Thus carbon composition and carbon film resistors have been replaced by metal film and metal oxide types, and thermionic valves have almost entirely been displaced by solid state devices.

In consequence the MTBF of a given item of equipment is some orders of magnitude larger than that of comparable equipment built in the 1950s.

However, the user of electronic apparatus has perceived a markedly smaller improvement than this, since the falling price of hardware due to integrated construction has enabled much more complex equipment to be produced economically.

Thus a typical instrument — for example an oscilloscope designed say 20 years ago — might use 7 or 8 valves while its modern counterpart might contain 60 or more transistors and several integrated circuits. Fortunately the increase in component MTBF is generally much greater than the increase in the number of components used, so the overall consequence is a more reliable product.

Inevitably, however, there is an unceasing demand for increasingly complicated equipment which embodies more functions and more automatic operation. Thus oscilloscopes at the upper end of the market include a microprocessor which displays the time and amplitude scales on the oscilloscope alongside the trace, and if required can measure pulse amplitude and rise time.

In order to increase the user's confidence in the data obtained, and to improve service to the user, much complex equipment now incorporates some degree of self-testing. Typically this ensures that after being switched on the equipment undertakes a series of tests of its functions, including where possible checking its calibration.

The test procedure can be more thorough, and may even include automatic adjustment, where a microprocessor is used for the normal control of the apparatus.

For example high performance tape recorders can be arranged to record a series of tones, firstly with varying bias, and then with fixed bias and varying equalisation. The data so recorded is then played back and the amplitudes are measured and stored. The built-in microprocessor can then determine the correct bias level and the appropriate degree of h.f. equalisation, and set up the recording circuits to provide this. The complete procedure can be concluded within a few seconds; the tape is then automatically rewound to the beginning and control of the recorder is handed over to the user.

In addition to the initial power-up check of apparatus provided by built-in test equipment, it can also give valuable diagnostic help to service engineers who may use the results to help fault finding and check repairs.

Although today's electronic designer has very reliable components at his disposal, and an ever widening range of integrated circuits, each new model of a familiar product becomes more complicated and has an increasingly large set of states into which it may be driven. Consequently the designer has to examine a much greater range of possibilities than did his predecessor working a few decades earlier, and the search for a reliable product becomes a longer and more arduous one.

We must therefore conclude that in all probability tomorrow's reliability engineer will have more difficult problems to solve and more complex systems to analyse than his present-day counterpart, and he will probably need all the help from computers and reliability data banks that he can get in order to meet the demands made of him.

PROBLEMS

1. One stage in a capacitance coupled amplifier consists of a transistor in the common-emitter configuration, with a collector load resistor R_1 connected to a 12 V supply, and a bias resistor R_2 = connected between base and collector. The emitter lead is earthed. R_1 and R_2 have nominal values of $1\,k\Omega$ and $47\,k\Omega$ respectively, with 10 per cent tolerance, and the transistor current gain h_{fe} lies within the limits of 20–70. The normal base-emitter voltage can be taken as 0·7 V, and the saturation voltage as V_{ce} = 0·4 V.

Calculate

1. The maximum sinusoidal output voltage which can be handled without clipping, under the worst-case conditions.
2. The maximum power dissipated in the transistor under the worst-case conditions.

[1. 2·14 V rms. (*Note*. It is necessary to calculate both V_c maximum and V_c minimum.)
2. 39·2 mW. (*Note*. This dissipation occurs for a particular value of R_1, in combination with a range of values of R_2 and h_{fe}.)]

2. Three amplifiers are used in parallel with the circuit shown in figure 5.2 to feed a $600\,\Omega$ load. The amplifiers are initially assembled with the diodes D omitted. The voltage feedback from the collector of the output transistor is found to reduce the voltage gain of each amplifier by 15 dB.

The input circuits of the amplifiers are then adjusted so that exactly equal voltages are produced at the outputs of the three amplifiers, and the diodes D are connected. The three amplifiers will then share the load equally. By what fraction will the output voltage be reduced if two of the amplifiers fail, causing the output transistor to become cut-off?

Assume that the incremental resistance of the diode D is $100\,\Omega$ when conducting.   [22·6 per cent]

3. In a double two-out-of-three voting arrangement as described in

section 5.17 show that, neglecting the failure probability of the voting circuits, the system reliability for a given operating period is

$$R = 1 - 27p^4$$

where p is the failure probability of one of the nine sensing units (assumed to be identical) in the same period.

Bibliography

Anderson, R. T., *Reliability Design Handbook* (IIT Research Institute, RADC, Griffiss Air Force Base, New York, 1976)

ARINC Research Corporation, *Reliability Engineering* (Prentice-Hall, Englewood Cliffs, N.J., 1964)

Barlow, R. E., and Proschan, F., *Mathematical Theory of Reliability* (Wiley, New York, 1965)

Barlow, R. E., and Proschan, F., *Statistical Theory of Reliability and Life Testing Probability Models* (Hold, C. HR & W, New York, 1975.)

Bazovsky, I., *Reliability Theory and Practice* (Prentice-Hall, Englewood Cliffs, N.J., 1961)

Beasock, J. V., *Discrete Semiconductor Reliability (DSR-3)* (Reliability Analysis Centre, RADC, Griffiss Air Force Base, New York, 1979)

Blanchard, B., and Lowery, E., *Maintainability, Principles and Practice* (McGraw-Hill, New York, 1969)

Bourne, A. J., and Green, A. E., *Reliability Technology* (Wiley Interscience, New York, 1972)

Brook, R. H. W., *Reliability Concepts in Engineering Manufacture* (Wiley, New York, 1972)

Brown, D. B., *Systems Analysis and Design for Safety* (Prentice-Hall, Englewood Cliffs, N.J., 1976)

Calabro, S. R., *Reliability Principles and Practice* (McGraw-Hill, New York, 1962)

Carter, A. C. S., *Mechanical Reliability* (Halsted Press, New York, 1973)

Cunningham, C. E., and Cox, W., *Applied Maintainability Engineering* (Wiley, New York, 1972)

Dijkstra, W. D., *A Discipline of Programming* (Prentice-Hall, Englewood Cliffs, N.J., 1973)

Dummer, G. W. A., and Winton, R. C., *An Elementary Guide to Reliability* (Pergamon, Oxford, 1965)

Dummer, G., and Griffin, N. B., *Electronic Reliability: Calculation and Design* (Pergamon, Oxford, 1966)

Flint, S. J., *Hybrid Circuit Data* (Reliability Analysis Centre, RADC, Griffiss Air Force Base, New York, 1979)

Halstead, N. H., *Elements of Software Science* (Elsevier–North Holland Inc., New York, 1977)

Henley, E. J., and Lynn, J. W. (ed.) *Generic Techniques in Systems Reliability Assessment* (Noordhoff International Publishing, Reading, Mass., 1976)

Hetzel, W. C. (ed.), *Program Test Methods* (Prentice-Hall, Englewood Cliffs, N.J., 1972)

I.E.C. Publication No. 605 – *Equipment Reliability Testing* (IEC, Geneva, 1978) (similar to MIL-STD-781)

Ireson, W. G., *Reliability Handbook* (McGraw-Hill, New York, 1966)

Jowett, C. E., *Reliability of Electronic Components* (Iliffe, London, 1966)

Jowett, C. E., *Reliable Electronic Assembly Production* (Tab Books, Blue Ridge Summit, Pa., 1971)

Kapur, K. C., and Lamberson, L. R., *Reliability in Engineering Design* (Wiley, New York, 1977)

Klein, M. R., *MDR-13 Memory/L.S.I. Data* (Reliability Analysis Centre, RADC, Griffiss Air Force Base, New York, 1979)

Kopetz, H., *Software Reliability* (Macmillan, London, 1979)

Locks, M. O., *Reliability, Maintainability and Availability Assessment* (Hayden, Rochelle Park, N.J., 1973)

Mann, W. C., and Wilcox, R. H., *Redundancy Techniques for Computing Systems* (Spartan, Washington, D.C., 1962)

Miller, I., and Freund, J. E., *Probability and Statistics for Engineers* (Prentice-Hall, Englewood Cliffs, N.J., 1977)

Myers, G. J., *Software Reliability: Principles and Practice* (Wiley, New York, 1976)

Nicholls, D. B., *MDR-12 Digital Failure Rate Data* (RADC, Griffiss Air Force Base, New York, 1979)

Peterson, W. W., and Weldon, E. J., *Error Correcting Codes* (MIT Press, Cambridge, Mass., 1971)

Pierce, W. H., *Failure-Tolerant Computer Design* (Academic Press, New York, 1965)

Radio Technical Commission for Aeronautics, *Airborne Electronics and Electrical Equipment Reliability* (RTCA, Washington, D.C., 1977)

Scheaffer, R. L., and Mendenhall, W., *Introduction to Probability: Theory and Applications* (Wadsworth, Belmont, Calif., 1975)

Shooman, M., *Probabilistic Reliability: An Engineering Approach* (McGraw-Hill, New York, 1968)

Smith, O., *Introduction to Reliability in Design* (McGraw-Hill, New York, 1976)

Smith, C. S., *Quality and Reliability* (Pitman, London, 1969)

Smith, C. J., *Reliability Engineering* (Barnes & Noble, Scranton, Pa., 1972)

Smith, D. J., and Babb, A. H., *Maintainability Engineering* (Pitman, London, 1973)

Swain, A. D., *The Human Element in System Safety* (Incomtech House, Camberley, UK, 1974)

U.S. Dept. of Defense, *Reliability Prediction of Electronic Equipment MIL-HDBK-217B* (U.S. Dept. of Defense, 1975)

Wakerley, J. F., *Error Detecting Codes, self-checking circuits and applications* (Elsevier–North Holland, New York, 1978)

Zelen, M. S., *Statistical Theory of Reliability* (Wisconsin Press, Madison, Wis., 1963)

Periodicals

I.E.E.E. Transactions on Reliability (I.E.E.E.)
Microelectronics and Reliability (Pergamon Press)

Standards, etc.

I.E.C. Recommendations

68–1 and 68–2 (1968): Basic Environmental Testing Procedures for Electronic Components and Electronic Equipment

319 (1970): Presentation of Reliability Data on Electronic Components (or parts)

382 (1971): Guide for the Collection of Reliability, Availability and Maintainability Data from Field Performance of Electronic Items.

British Standards

BS 4200: Parts 1–8   Guide on the reliability of electronic equipment and parts used therein

BS 4891: 1972   A guide to quality assurance

BS 5179: Parts 1–3   Guide to the operation and evaluation of quality
   assurance systems
BS 9000: Parts 1–2   General requirements for electronic components
   of assessed quality
BS 9003: 1977   Requirements for the manufacture of electronic
   components of assessed quality intended for long life applications
BS 9010–9765   Specifications for particular types of components of
   assessed quality.
E9000–E9377   Specifications for electronic components of assessed
   quality harmonized with the CENELEC Electronic Components
   Committee system

British Standard Codes of Practice
CP 1003: Parts 1–3   Electrical apparatus and associated equipment
   for use in explosive atmospheres of gas or vapour other than mining
   applications
CP 1013: 1965   Earthing
CP 1016: Parts 1–2   The use of semiconductor devices

British Standards Handbooks
PD 6436   A guide to the BS 9000 system
PD 9002: Parts 1–9   BS 9000 component selection guide
PD 9004   BS 9000, CECC and IECQA–UK administrative guide.
   Procedures for the National implementation of quality assessment
   systems for electronic components

Index

175